It's another star from the CGP galaxy...

Here's the thing: you can't cut corners when it comes to Grade 9-1 GCSE Physics. You really have to practise until you're 100% confident about every topic.

That's where this indispensable CGP book comes in. It's bursting with questions just like the ones you'll face in the real exams, including those tricky required practicals.

And since you'll be tested on a wide range of topics in the real exams, we've also included a section of mixed questions to keep you on your toes!

CGP — still the best! ☺

Our sole aim here at CGP is to produce the highest quality books — carefully written, immaculately presented and dangerously close to being funny.

Then we work our socks off to get them out to you — at the cheapest possible prices.

Contents

☑ Use the tick boxes to check off the topics you've completed.

Topic 7 — Magnetism and Electromagnetism

Topic 8 — Space Physics

Mixed Questions

Published by CGP

Editors:
Emily Garrett, Sharon Keeley-Holden, Duncan Lindsay, Frances Rooney, Charlotte Whiteley,
Sarah Williams and Jonathan Wray.

Contributors:
Mark A. Edwards, Daniel Limb, Barbara Mascetti, Brian Mills and Jonathan Schofield.

With thanks to Ian Francis, Rachael Marshall and Karen Wells for the proofreading.

Data on page 80 contains public sector information licensed under the Open Government Licence v3.0.
http://www.nationalarchives.gov.uk/doc/open-government-licence/version/3/

Clipart from Corel®
Printed by Elanders Ltd, Newcastle upon Tyne

Based on the classic CGP style created by Richard Parsons.

How to Use This Book

- Hold the book <u>upright</u>, approximately <u>50 cm</u> from your face, ensuring that the text looks like <u>this</u>, not ᔕ⊥Ɥ⊥. Alternatively, place the book on a <u>horizontal</u> surface (e.g. a table or desk) and sit adjacent to the book, at a distance which doesn't make the text too small to read.

- In case of emergency, press the two halves of the book together <u>firmly</u> in order to close.

- Before attempting to use this book, familiarise yourself with the following <u>safety information</u>:

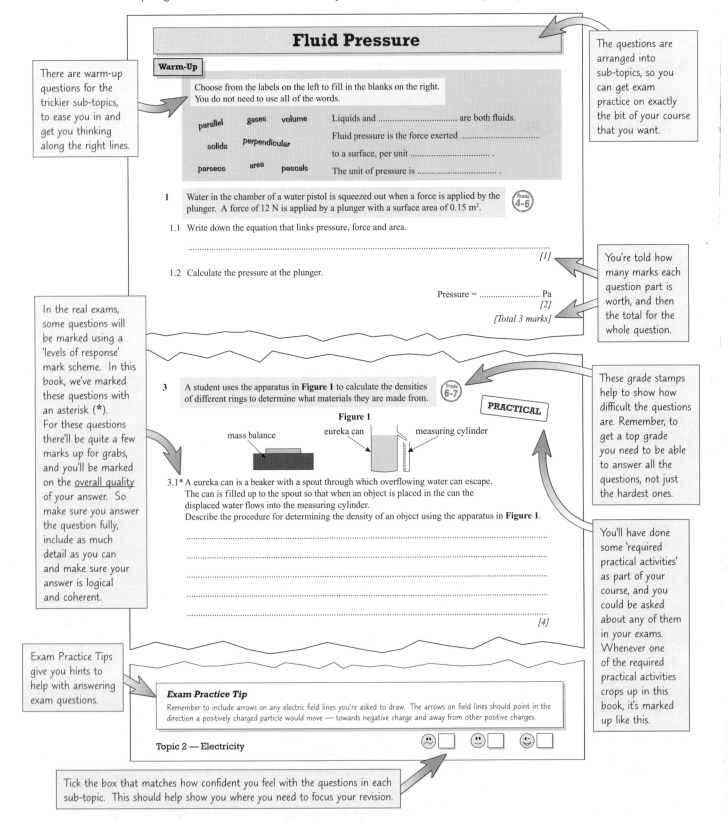

There are warm-up questions for the trickier sub-topics, to ease you in and get you thinking along the right lines.

The questions are arranged into sub-topics, so you can get exam practice on exactly the bit of your course that you want.

Fluid Pressure

Warm-Up

Choose from the labels on the left to fill in the blanks on the right. You do not need to use all of the words.

parallel   gases   volume

solids   perpendicular

parsecs   area   pascals

Liquids and .... are both fluids.

Fluid pressure is the force exerted ....

to a surface, per unit ....

The unit of pressure is ....

1   Water in the chamber of a water pistol is squeezed out when a force is applied by the plunger. A force of 12 N is applied by a plunger with a surface area of 0.15 m². (Grade 4-6)

1.1 Write down the equation that links pressure, force and area.

....
[1]

1.2 Calculate the pressure at the plunger.

Pressure = ... Pa
[2]
[Total 3 marks]

You're told how many marks each question part is worth, and then the total for the whole question.

In the real exams, some questions will be marked using a 'levels of response' mark scheme. In this book, we've marked these questions with an asterisk (*). For these questions there'll be quite a few marks up for grabs, and you'll be marked on the <u>overall quality</u> of your answer. So make sure you answer the question fully, include as much detail as you can and make sure your answer is logical and coherent.

3   A student uses the apparatus in **Figure 1** to calculate the densities of different rings to determine what materials they are made from. (Grade 6-7)

PRACTICAL

Figure 1

mass balance   eureka can   measuring cylinder

3.1* A eureka can is a beaker with a spout through which overflowing water can escape. The can is filled up to the spout so that when an object is placed in the can the displaced water flows into the measuring cylinder. Describe the procedure for determining the density of an object using the apparatus in **Figure 1**.

....
....
....
....
....
[4]

These grade stamps help to show how difficult the questions are. Remember, to get a top grade you need to be able to answer all the questions, not just the hardest ones.

You'll have done some 'required practical activities' as part of your course, and you could be asked about any of them in your exams. Whenever one of the required practical activities crops up in this book, it's marked up like this.

Exam Practice Tips give you hints to help with answering exam questions.

Exam Practice Tip

Remember to include arrows on any electric field lines you're asked to draw. The arrows on field lines should point in the direction a positively charged particle would move — towards negative charge and away from other positive charges.

Topic 2 — Electricity   ☹ ☐   ☺ ☐   ☺ ☐

Tick the box that matches how confident you feel with the questions in each sub-topic. This should help show you where you need to focus your revision.

- There's also an Equations List at the back of this book — you'll be given these equations in your exam. You can look up equations on this list to help you answer some of the questions in this book.

Specific Heat Capacity

Which of the following is the correct definition of specific heat capacity? Tick **one** box.

The energy transferred when an object is burnt. ☐

The maximum amount of energy an object can store before it melts. ☐

The energy needed to raise 1 kg of a substance by 10 °C. ☐

The energy needed to raise 1 kg of a substance by 1 °C. ☐

PRACTICAL

1 **Figure 1** shows the apparatus used by a student to investigate the specific heat capacities of various liquids. She measured out 0.30 kg of each substance, then supplied 15 kJ of energy to each sample using an immersion heater. She then recorded her results, shown in **Table 1**.

Grade 6-7

Figure 1

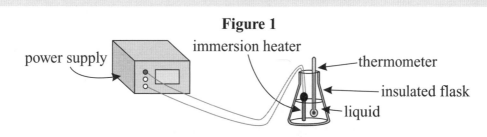

power supply · immersion heater · thermometer · insulated flask · liquid

Table 1

Liquid	Mass (kg)	Temperature change (°C)	Specific heat capacity (J/kg °C)
A	0.30	12	4200
B	0.30	23	2200
C	0.30	25	....

1.1 Complete **Table 1** by calculating the specific heat capacity of Liquid C.
Use an equation from the Equations List.

[3]

1.2 Describe the energy transfers that occur when a liquid is heated using the equipment in **Figure 1**.

....

....

....

....

[4]

[Total 7 marks]

Exam Practice Tip

You may be asked about experiments you've never seen before in an exam, but don't panic. Take your time to read the experiment carefully and work out what's going on before attempting any questions to get full marks.

😐 ☐   🙂 ☐   😊 ☐

4

Conservation of Energy and Power

Choose from the words on the left to fill in the blanks in the sentences on the right. You do not need to use all of the words.

joules   work done

total   minimum

energy lost   rate of   watts

Power is the .... energy transfer or

.... It is measured in ....

1   An electric fan wastes some energy by transferring it to the thermal energy stores of its surroundings. Describe what is meant by 'wasted energy'. *(Grade 4-6)*

....

....

....

[Total 1 mark]

2   **Figure 1** shows a rechargeable battery-powered shaver. The shaver transfers some energy to useful energy stores and wastes some energy. *(Grade 4-6)*

Figure 1

2.1 Which statements about energy are false? Tick **two** boxes.

Energy can be transferred usefully. ☐

Energy can be created. ☐

Energy can be stored. ☐

Energy can be dissipated. ☐

Energy can be destroyed. ☐

[1]

2.2 Give **one** example of a useful energy store and **one** of a wasted energy store that the shaver transfers energy to.

Useful energy store: ....

Wasted energy store: ....

[2]

2.3 Describe what effect increasing the power of the shaver would have on the shaver's battery life.

....

[1]

[Total 4 marks]

Topic 1 — Energy

3 A student is investigating the insulating properties of various materials. He surrounds a beaker of water with each material, before heating the water using an electric immersion heater with a constant power of 35 W.

(Grade 6-7)

3.1 Write down the equation that links power, work done and time.

....

[1]

3.2 Calculate the work done by the immersion heater when it is operated for 600 s.

Work done = ... J

[2]

3.3 Whilst investigating the insulating properties of cotton wool, the student forgets to measure the time that he leaves the immersion heater on for. Calculate the time that the heater was on for, if it transferred 16 800 J of energy to the system.

Time = ... s

[2]

[Total 5 marks]

4 A car contains a worn out engine with a power of 32 000 W. The car takes 9.0 s to accelerate from rest to 15 m/s. A mechanic replaces the engine with a more powerful but otherwise identical one. The new engine has a power of 62 000 W.

(Grade 7-9)

4.1 Explain how the new engine will affect the time it takes for the car to accelerate from rest to 15 m/s.

....

....

....

....

[3]

4.2 Calculate how long it will take for the car to accelerate to 15 m/s now. You can assume that the total amount of energy wasted whilst the car is accelerating is the same for both engines.

Time = ... s

[4]

[Total 7 marks]

Exam Practice Tip

For really big powers, you might see the unit kW, which stands for kilowatt. Don't let this put you off though, you just need to remember that 1000 W = 1 kW. You might see this in a few other units too, for example 1000 m = 1 km.

☹ ☐   😐 ☐   🙂 ☐

Topic 1 — Energy

6

Conduction and Convection

1   Use words from the box below to complete the passage. You can only use each word **once** and you do not need to use all of the words.

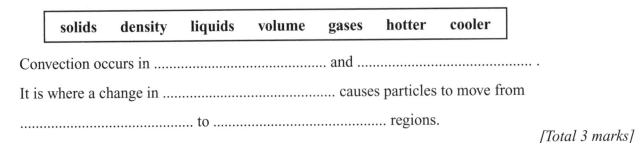

| solids | density | liquids | volume | gases | hotter | cooler |

Convection occurs in .... and ....

It is where a change in .... causes particles to move from

.... to .... regions.

[Total 3 marks]

2   A student uses the apparatus in **Figure 1** to investigate conduction. She heats blocks of different materials, shown in **Figure 2**, and uses a stopwatch to measure the time it takes for the upper surface of the block to increase in temperature by 2 °C.

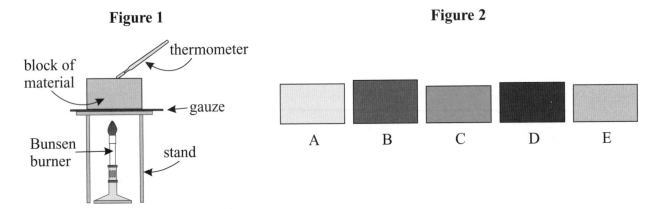

Figure 1

Figure 2

2.1 Suggest **one** way to improve the student's experiment.

....

[1]

2.2 Describe, in terms of particles, the energy transfers that take place within a block as it is heated.

....

....

....

....

[3]

Table 1

Block	A	B	C	D	E
Time taken (s)	83	37	74	97	86

2.3 **Table 1** shows the student's results.
     Suggest a conclusion you can make about block B, compared to the other blocks.

....

[1]

[Total 5 marks]

Topic 1 — Energy

Reducing Unwanted Energy Transfers

Which of the following options would reduce the air resistance acting on a cyclist?
Circle **one** box.

| Wearing clothes that are more thermally insulating | Lubricating the bicycle's wheels | Changing the colour of his clothes | Wearing a more streamlined helmet |

1   Which of the following statements are true? Tick **two** boxes.

The thickness of a house's walls does not affect the rate at which it loses energy. ☐

Thicker walls decrease the rate of energy lost from a house. ☐

Thicker walls increase the rate of energy lost from a house. ☐

Bricks with a higher thermal conductivity transfer energy at a faster rate. ☐

[Total 1 mark]

2   **Figure 1** shows a thermal image of a house. Different parts of
the outside of the house are at different temperatures. The owner
wants to keep the inside of the house as warm as possible.

Figure 1

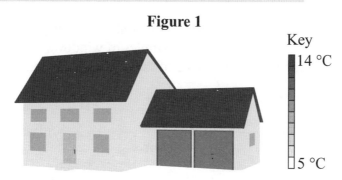

Key

14 °C

5 °C

2.1 Suggest where the highest rate of unwanted energy transfer occurs in the house.

....

[1]

2.2 Suggest **one** way to reduce this unwanted energy transfer.

....

[1]

2.3 A second house is tested and it is found that the majority of its unwanted energy transfers occur
around the doors and windows. Suggest **two** ways to reduce these energy transfers.

1. ....

2. ....

[2]

[Total 4 marks]

8

3 **Figure 2** shows an old-fashioned well. The handle is turned, which rotates the axle. This causes the rope attached to the bucket to wrap around the axle, raising the bucket from the well.

Grade 6-7

Figure 2

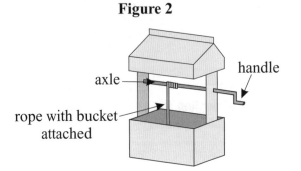

A student measures the time taken to raise a bucket of water from the well. After lubricating the axle of the well, the student repeats the test and finds the time taken is shorter. Explain why.

....

....

....

....

....

[Total 3 marks]

PRACTICAL

4 A student investigates which type of window is the best at reducing unwanted energy transfers. The student places different samples of windows on a hot plate and measures how long it takes for the top surface of the window sample to reach 30 °C.

Grade 7-9

Figure 3

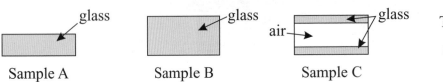

Thermal conductivity

glass = 0.2 W/mK

air = 0.03 W/mK

Sample A   Sample B   Sample C

Figure 3 shows the cross-sections of each window sample. Rank them from best to worst for reducing unwanted energy transfers from a house and explain your choices.

Best: ....   Second best: ....   Worst: ....

....

....

....

....

....

[Total 5 marks]

Efficiency

1   20 kJ of energy is transferred to a mobile phone battery to fully charge it once it has lost all charge. It transfers 16 kJ of useful energy during use until it needs to be recharged.   (Grade 4-6)

1.1   Write down the equation that links efficiency, total input energy transfer and useful output energy transfer.

....

[1]

1.2   Calculate the efficiency of the battery.

Efficiency = ...

[2]

[Total 3 marks]

2   An electric motor has a useful power output of 57 W and an efficiency of 75%.
Calculate the total power input for the motor. Use an equation from the Equations List.   (Grade 4-6)

Input power = ... W

[Total 3 marks]

3   A student investigates the efficiency of a scale model of an electricity generating wind turbine using the equipment in **Figure 1**.   (Grade 6-7)

The student changes the number of sails on the turbine and measures the power output from the turbine's generator. The air blower is supplied with 533 W and has an efficiency of 0.62.

Figure 1

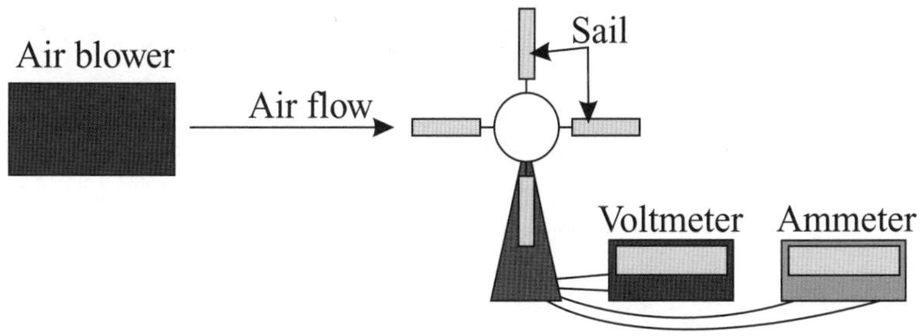

3.1   When using two sails, the efficiency of the turbine was 13%. Calculate the power generated.
Give your answer to 2 significant figures.

Output power = ... W

[4]

3.2   Suggest **two** ways the student could increase the efficiency of the turbine.

1. ....

2. ....

[2]

[Total 6 marks]

Energy Resources and Their Uses

Write the resources below in the correct place in the table to show whether they are renewable or non-renewable energy resources.

bio-fuel   oil

coal

hydroelectricity

solar

wind

nuclear fuel

tidal   geothermal

wave power   gas

Renewable	Non-renewable

1 Describe the difference between renewable and non-renewable energy resources. *(Grade 4-6)*

....

....

[Total 2 marks]

2 Most cars run on petrol or diesel, which are both derived from fossil fuels. *(Grade 4-6)*

2.1 Name the **three** fossil fuels.

....

[1]

2.2 Give **two** other everyday uses for fossil fuels.

1. ....

2. ....

[2]

2.3 Some modern cars are made to run on bio-fuels. What are bio-fuels?

....

....

[1]

2.4 Suggest **one** reason why car manufacturers are developing cars that run on alternative fuels to petrol and diesel.

....

....

[1]

[Total 5 marks]

3 A UK university is considering ways to reduce their energy bills. They are considering building either a single wind turbine nearby, or installing solar panels on top of their buildings. By commenting on the change of seasons throughout the year, suggest why the university may decide to install both wind turbines and solar panels.

Grade 6-7

....

....

....

....

[Total 5 marks]

4 An energy provider is looking to replace their old fossil fuel power plant. They are eligible for a government grant, so the initial building costs are negligible.

Grade 7-9

4.1* The energy provider is interested in building a power plant that uses renewable energy resources. They have narrowed their choice to either a hydroelectric power plant or a tidal barrage. Compare generating electricity from hydroelectricity and tides, commenting on their reliability and their impact on the environment.

....

....

....

....

....

....

[4]

4.2* An alternative is replacing the old power plant with a new power plant that is run on fossil fuels. Discuss the advantages and disadvantages of using fossil fuels to generate electricity.

....

....

....

....

....

....

....

....

....

....

[6]

[Total 10 marks]

Trends in Energy Resource Use

1   **Figure 1** shows the energy resources used to generate electricity in a country. *Grade 4-6*

Figure 1

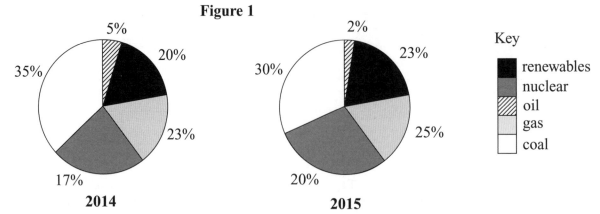

2014

2015

Key
- renewables
- nuclear
- oil
- gas
- coal

1.1   Determine what percentage of the country's electricity was generated by fossil fuels in 2014.

...........................%
[2]

1.2   Suggest **one** trend you can determine from the graphs in **Figure 1**.

....
[1]

[Total 3 marks]

2*   In the UK, the use of renewable energy resources is increasing, but many say it is not increasing at a fast enough rate. Suggest reasons for this increase in the use of renewable energy resources. Suggest and explain the factors that may affect the speed at which we use more renewable energy resources. *Grade 7-9*

....

....

....

....

....

....

....

....

....

....

....
[Total 6 marks]

Current and Circuit Symbols

Warm-Up

Draw lines to match circuit symbols A-D to their correct names.

A   B   C   D

switch   filament lamp   fuse   cell

1   **Figure 1** shows a simple circuit, featuring a 10 Ω resistor.   (Grade 4-6)

Figure 1

$10\ \Omega$

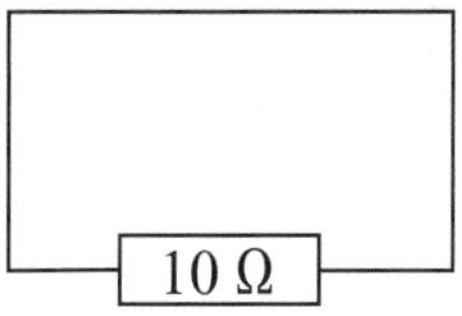

1.1   Explain why there is no current in the circuit.

....

[1]

1.2   Use a word from the following list to complete the sentence below:

charge	potential difference	resistance	frequency

Current is the rate of flow of ....

[1]

[Total 2 marks]

2   **Figure 2** shows two ammeters, A_1 and A_2, in a circuit. The reading on A_1 is 0.5 A.   (Grade 4-6)

Figure 2

2 V

A_1 A_2

X

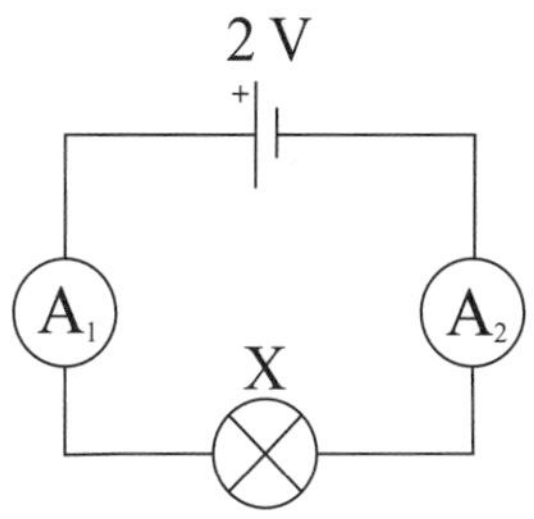

2.1   What is the reading on A_2?

Current = ... A

[1]

2.2   State the equation that links electric charge, time and current.

....

[1]

2.3   Calculate the charge that flows through component **X** in 2 minutes. Give the unit in your answer.

Charge = ... Unit = ...

[3]

[Total 5 marks]

Exam Practice Tip

You'll need to know the basics of how circuits work and the different circuit symbols for most electricity questions.

Resistance and V = IR

1 A current of 3 A flows through a 6 Ω resistor.
   Calculate the potential difference across the resistor.

Potential Difference = ...V

[Total 2 marks]

PRACTICAL

2 A student investigated how the resistance of a piece of wire depends on its length. Grade 6-7
   The circuit she used is shown in **Figure 1**. Her results are displayed in **Table 1**.

Figure 1

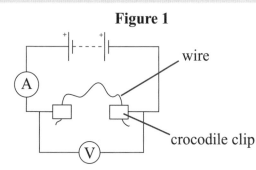

Table 1

Length / cm	Resistance / Ω
10	0.6
20	1.3
30	1.7
40	2.4
50	3.0

2.1 Describe how the student could have used the apparatus in **Figure 1** to obtain the results in **Table 1**.

   ....

   ....

   [2]

2.2 Plot a graph of the data in **Table 1** on the grid shown in **Figure 2**.
   Label the axes correctly. Draw a line of best fit on the graph.

Figure 2

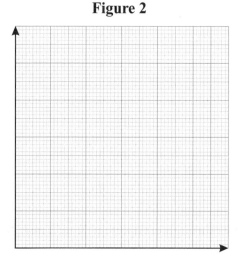

   [4]

2.3 State **one** conclusion the student can make about the relationship between the resistance of a wire
   and its length. Explain how **Figure 2** shows this.

   ....

   ....

   [2]

   [Total 8 marks]

Resistance and I-V Characteristics

1   **Figure 1** shows some graphs of current against potential difference.   Grade 4-6

1.1 Tick the box below the correct graph for a resistor at constant temperature.

Figure 1

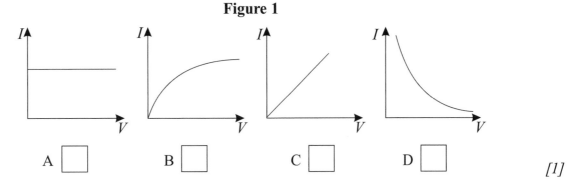

A ☐   B ☐   C ☐   D ☐   *[1]*

1.2 Name the type of graph shown in **Figure 1**.

....

[1]

1.3 Use words from the following list to complete the sentences below:

| linear | non-linear | non-ohmic | ohmic |

A resistor at a constant temperature is an example of a(n) ....

conductor. It is also an example of a(n) .... component.

[2]

[Total 4 marks]

2   This question is about diodes.   Grade 6-7

2.1 Draw the standard circuit symbol for a diode.

[1]

2.2 An old name for a diode is a valve. A valve in a bicycle pump only lets air flow through
it in one direction. In what way do diodes behave in a similar way to valves?

....

[1]

2.3 A student measured the resistance of a diode using an electric circuit. He found the resistance
to be 0.02 Ω. The next day he measured the diode again. This time he measured the resistance
to be 100 MΩ. Suggest why the student's measurements were so different.
You may assume that the circuit is working perfectly on both occasions.

....

....

[2]

[Total 4 marks]

PRACTICAL

3 A student used the circuit in **Figure 2** to find the *I-V* characteristic of a filament lamp.

Figure 2

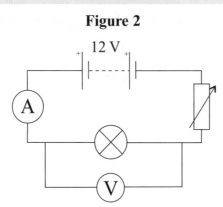

3.1 Explain the purpose of the variable resistor in the circuit.

....

....

[2]

3.2 The student obtained the graph displayed in **Figure 3**.
Use the graph to find the resistance of the lamp at 3 A.

Figure 3

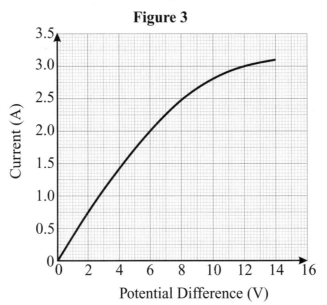

Resistance = ... Ω

[4]

3.3 What does the graph tell you about the lamp's resistance as the current increases?
Explain why the resistance behaves in this way.

....

....

[2]

3.4 The student states that the lamp behaves as an ohmic conductor up to a potential difference of
approximately 3.5 V. Explain what has led the student to this conclusion.

....

....

[2]

[Total 10 marks]

Topic 2 — Electricity

Circuit Devices

1   A student wants to measure the resistance of a light dependent resistor.

1.1   Draw a circuit diagram (including an ammeter and a voltmeter)
that can be used to measure the resistance of an LDR.

[3]

The resistance of an LDR changes depending on its surroundings.

1.2   State what happens to the resistance of an LDR as the surrounding light intensity increases.

....

[1]

1.3   Give **one** example of a device that uses a light dependent resistor.

....

[1]

[Total 5 marks]

2   **Figure 1** shows a circuit that can be used for a light
that lights up when the surface of a cooker is hot.
Describe how the circuit works.

Figure 1

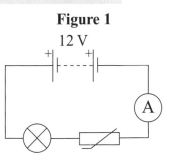

....

....

....

....

....

[Total 4 marks]

Exam Practice Tip

You may be asked to interpret the resistance graph of an LDR or thermistor. Remember the graph starts off steep and then
levels out. So for an LDR at low light intensities, a small change in light intensity will cause a large change in resistance.

Series Circuits

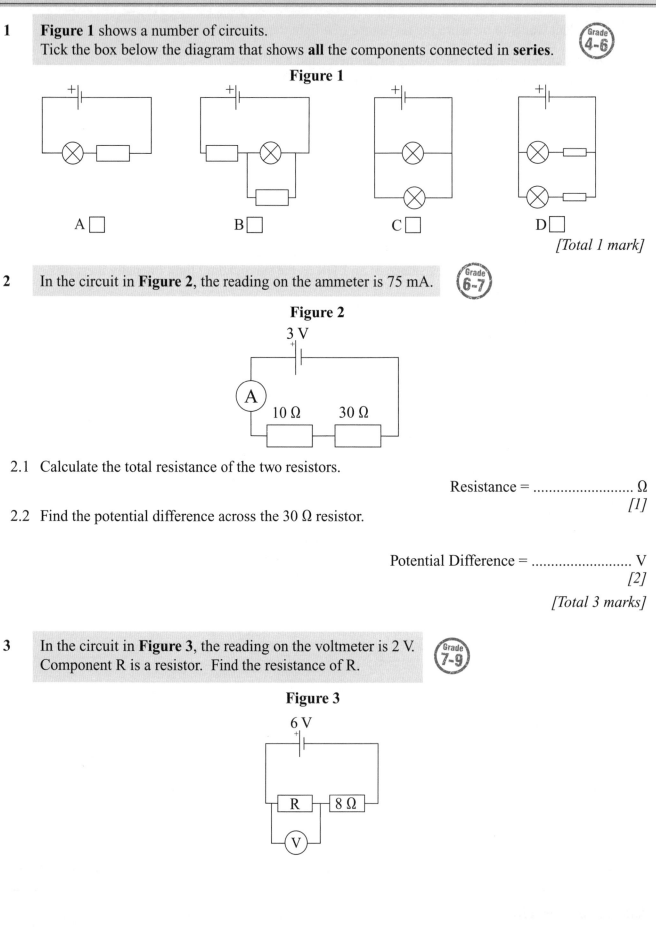

1   **Figure 1** shows a number of circuits.
Tick the box below the diagram that shows **all** the components connected in **series**. Grade 4-6

Figure 1

A ☐   B ☐   C ☐   D ☐

[Total 1 mark]

2   In the circuit in **Figure 2**, the reading on the ammeter is 75 mA. Grade 6-7

Figure 2

3 V

A

10 Ω   30 Ω

2.1   Calculate the total resistance of the two resistors.

Resistance = ... Ω
[1]

2.2   Find the potential difference across the 30 Ω resistor.

Potential Difference = ... V
[2]

[Total 3 marks]

3   In the circuit in **Figure 3**, the reading on the voltmeter is 2 V. Grade 7-9
Component R is a resistor. Find the resistance of R.

Figure 3

6 V

R   8 Ω

V

Resistance = ... Ω
[Total 5 marks]

Topic 2 — Electricity   ☹ ☐   😐 ☐   ☺ ☐

Parallel Circuits

1 Draw a circuit diagram consisting of a cell and two filament lamps connected in parallel. [Grade 4-6]

[Total 1 mark]

2 **Figure 1** shows a circuit with a 6 V supply and 4 Ω and 12 Ω resistors connected in parallel. There are also three ammeters and two voltmeters in the circuit. [Grade 6-7]

Figure 1

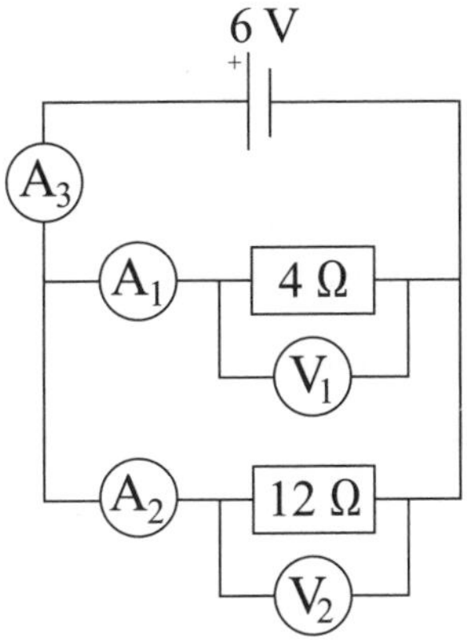

2.1 Determine the readings on voltmeters V_1 and V_2.

Potential difference = ... V
[1]

2.2 Calculate the currents through A_1 and A_2.

A_1 current = ... A, A_2 current = ... A
[5]

2.3 Calculate the current from the supply as measured by A_3.

A_3 = ... A
[1]

[Total 7 marks]

3* Explain why adding resistors in series increases the total resistance, whilst adding resistors in parallel decreases the total resistance. [Grade 7-9]

....

....

....

....

....

....

....

[Total 6 marks]

Topic 2 — Electricity

Investigating Resistance

1 A student is investigating how adding identical fixed resistors in series affects the resistance of the circuit. **Figure 1** shows his results.

Figure 1

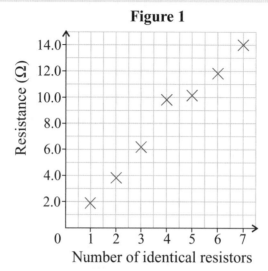

1.1 The student made a mistake when plotting his results. Draw a line of best fit for the student's data on **Figure 1**. Use this to predict the correct resistance for the incorrectly plotted result.

Resistance = ... Ω

[2]

1.2 The student repeats his experiment, this time using 1 Ω resistors.
Draw the predicted line of best fit for the results of this experiment on the axes in **Figure 1**.

[2]

[Total 4 marks]

2* A student wants to investigate how adding fixed resistors in parallel affects the overall resistance of a circuit. Describe an experiment the student could do to investigate this. You may draw a circuit diagram as part of your answer.

....

....

....

....

....

....

....

[Total 6 marks]

Electricity in the Home

Warm-Up

Use the words given to complete the sentences about the wires in three-core cables.

| green and yellow | 0 | 230 | brown |

The live wire is .... and is at a potential difference of .... V.

The earth wire is .... and is at a potential difference of .... V.

1   A toaster is connected to the mains electricity supply using a three-core cable.   *Grade 6-7*

1.1   State the frequency and potential difference of the UK mains supply.

....

[2]

1.2* The toaster cable has a fault such that the live wire is in electrical contact with the neutral wire. Explain why the toaster will not work while this fault remains.

....

....

....

....

[4]

[Total 6 marks]

2   The cable that connects an iron to the mains supply has become worn with use. There is no insulation covering part of the live wire. The iron is plugged in, but switched off.   *Grade 6-7*

2.1   State **two** purposes of the insulation that covers the live wire.

....

....

[2]

2.2   A man switches on the iron and touches the exposed live wire. He receives an electric shock. With reference to the electrical potential of the man, explain why he receives an electric shock.

....

....

....

[3]

2.3   The socket is switched off and the iron is unplugged. Explain whether there is still a danger of the man receiving an electric shock from the plug socket.

....

....

....

[3]

[Total 8 marks]

Power of Electrical Appliances

1   Use the correct words from the following list to complete the sentences below.   *Grade 4-6*

current	power	in total	per second	potential difference	safety

The .... of an appliance is the energy transferred ....

Energy is transferred because the .... does work against the appliance's resistance.

[Total 3 marks]

2   A child is playing with a toy car. The car is powered by a   *Grade 6-7*
battery and has two speed settings — fast and slow.

2.1 The child sets the speed to slow and drives the car for 20 seconds. The power of the
car at this speed is 50 W. Write down the formula that links energy, power and time.

....

[1]

2.2 Calculate the energy transferred by the car.

Energy transferred = ... J

[2]

2.3 The child now sets the speed to fast. The power of the car at this speed is 75 W.
Explain why the battery runs down more quickly when the car is set at a higher speed.

....

....

[2]

[Total 5 marks]

3   **Table 1** shows some data for two   *Grade 7-9*
different cycles of a washing machine.

Table 1

Cycle	Power	Time needed
Standard Mode	600 W	125 minutes
Economy Mode	400 W	160 minutes

3.1 Name the **two** main useful energy transfers that take place in the washing machine.

....

....

[3]

3.2 Calculate the work done by the washing machine per minute
when the machine is in Economy Mode.

Work done = ... J

[2]

3.3 Calculate the energy saved per cycle by using Economy Mode instead of Standard Mode.

Energy saved = ... J

[4]

[Total 9 marks]

More on Power

Use the words below to fill in the gaps in the passage about energy in a circuit.

work   resistance   energy   decreases

A power source supplies .... to a charge.

When a charge passes through a component with .... ,

it does .... , so the charge's energy ....

1   **Figure 1** shows a circuit. The reading on the voltmeter is 6 V
and the reading on the ammeter is 2 A. This means 2 coulombs
of electric charge pass through the ammeter every second.

(Grade 6-7)

Figure 1

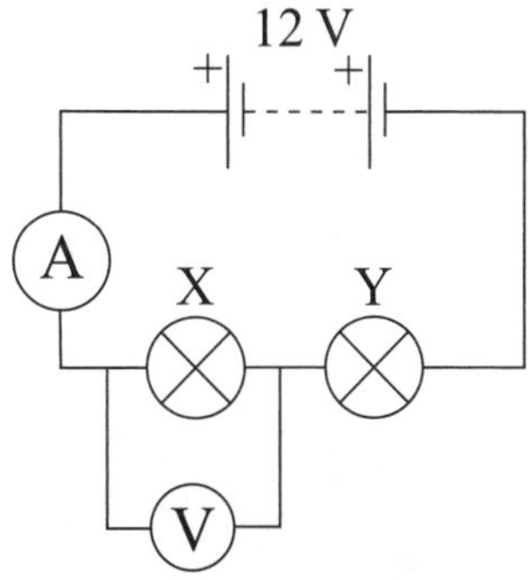

1.1 Write down the equation that links potential difference, charge and
energy transferred.

....
[1]

1.2 Calculate the energy transferred to lamp X when 2 C of charge
passes through it.

Energy transferred = ... J
[2]

1.3 Explain why multiplying the current through lamp X by the potential
difference across it will give you the same value as in 1.2.

....

....

....
[3]
[Total 6 marks]

2   Fans use a motor to turn a set of blades. *(Grade 7-9)*

2.1 A 75 W ceiling fan in an office is powered by the mains supply at 230 V.
Calculate the current supplied to the fan.

Current = ... A
[2]

2.2 A smaller fan on someone's desk runs from a computer's USB port.
It has a power of 2.5 W, and draws a current of 0.50 A. Calculate its resistance.

Resistance = ... Ω
[2]
[Total 4 marks]

The National Grid

1 The national grid uses **transformers** to transfer energy efficiently. (Grade 4-6)

1.1 Which **two** of the following quantities are changed by a transformer? (Assume the transformer is 100% efficient.) Put ticks in the boxes next to the correct answers.

☐ Power ☐ Potential Difference ☐ Current ☐ Resistance

[2]

1.2 Describe the difference in the function of a step-up and a step-down transformer.

....

[1]

[Total 3 marks]

2 **Figure 1** shows a diagram of part of the national grid which transfers energy from a power station to a home. (Grade 6-7)

Figure 1

Power Station → Transformer A → Power Cables → Transformer B → Home

2.1 What types of transformer are transformers A and B?

Transformer A = ...

Transformer B = ...

[2]

2.2* Explain how transformer A helps to improve the efficiency of the national grid.

....

....

....

....

....

....

[4]

2.3 Explain the purpose of transformer B.

....

....

[2]

[Total 8 marks]

Static Electricity

1   Two balloons are charged up and attached to a
     ceiling using thread, as shown in **Figure 1**.   (Grade 4-6)

Figure 1

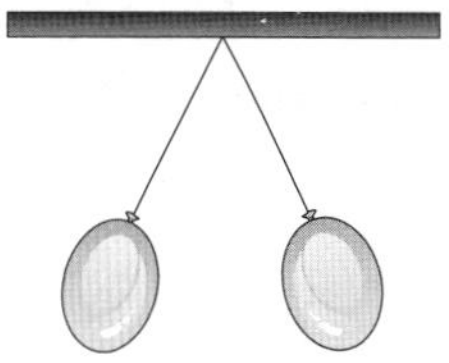

1.1 Using **Figure 1**, describe whether the static charges
     on the balloons are alike or opposite. Explain your answer.

     ....

     ....
                                                                                                              [2]

1.2 Suggest how the balloons may have been charged up.

     ....
                                                                                                              [1]
                                                                                                [Total 3 marks]

2   A student rubs a polythene rod with a dusting cloth. The rod becomes
     negatively charged and the dusting cloth becomes positively charged.   (Grade 6-7)

2.1 Describe what happens to the electrons as the polythene rod is rubbed.

     ....

     ....
                                                                                                              [2]

2.2 The rod is now suspended from a string tied around its centre. Describe how the student
     could use this set-up and the dusting cloth to show that opposite charges attract.

     ....

     ....
                                                                                                              [2]
                                                                                                [Total 4 marks]

3   A man walks up some carpeted stairs. The handrail is made of metal and is electrically
     connected to earth. When he puts his hand near the rail, there is a spark.   (Grade 7-9)

3.1 The carpet and the man's shoes rub together, making the man electrically charged.
     Explain why there is a spark between the man's hand and the rail.

     ....

     ....
                                                                                                              [2]

3.2 Given that the spark leapt from the man to the handrail, was the man positively charged
     or negatively charged? Explain your answer.

     ....

     ....
                                                                                                              [3]
                                                                                                [Total 5 marks]

Topic 2 — Electricity

Electric Fields

1   **Figure 1** shows a negatively charged sphere. (Grade 6-7)

1.1 Draw field lines on **Figure 1** to show the electric field around the sphere.

Figure 1

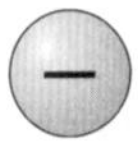

[1]

1.2 Explain what is meant by an electric field.

....

[1]

1.3 State what happens to the strength of the field as you move away from the charged sphere.

....

[1]

1.4 A second sphere is placed inside the electric field of the first sphere.
This new sphere does not experience an electric force. Suggest why this is the case.

....

[1]

[Total 4 marks]

2   **Figure 2** shows an electric field between two oppositely charged spheres.
An air particle is also shown (not to scale). The air particle is made up
of both positive and negative charges. It experiences a non-contact force. (Grade 6-7)

Figure 2

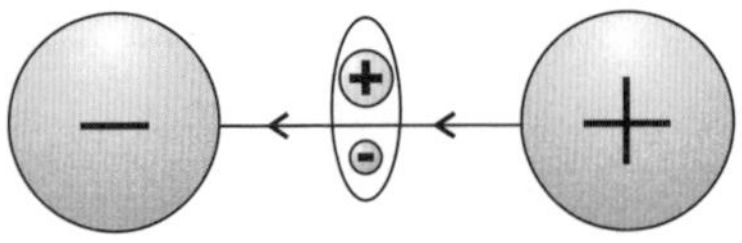

2.1 The potential difference between the spheres is increased.
State what happens to the size of the force experienced by the air particle.

....

[1]

2.2 When the potential difference is high enough, the air particle begins to break apart.
Explain why this happens.

....

....

....

[3]

[Total 4 marks]

Exam Practice Tip

Remember to include arrows on any electric field lines you're asked to draw. The arrows on field lines should point in the
direction a positively charged particle would move — towards negative charge and away from other positive charges.

Density of Materials

Warm-Up

The images below show the particles in a substance when it is in three different states of matter.
Label each image to show whether the substance is a solid, a liquid or a gas.

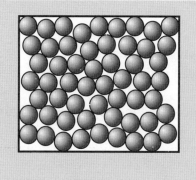 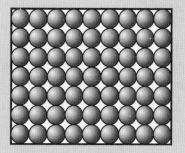 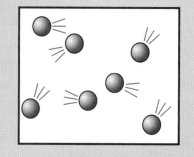

....

1   A 0.5 m³ block of tungsten has a mass of 10 000 kg.   (Grade 4-6)

1.1   Write down the equation that links density, mass and volume.

....

[1]

1.2   Calculate the density of tungsten.

Density = ... kg/m³

[2]

1.3   Calculate the mass of a 0.02 m³ sample cut from the tungsten block.

Mass = ... kg

[3]

[Total 6 marks]

2   Eric notices that ice cubes float when he puts them into a glass of water.
This is because ice is less dense than liquid water. Explain what this
suggests about the arrangement of the water molecules in each state.   (Grade 4-6)

....

....

....

....

[Total 2 marks]

3 A student uses the apparatus in **Figure 1** to calculate the densities of different rings to determine what materials they are made from.

Grade 6-7

Figure 1

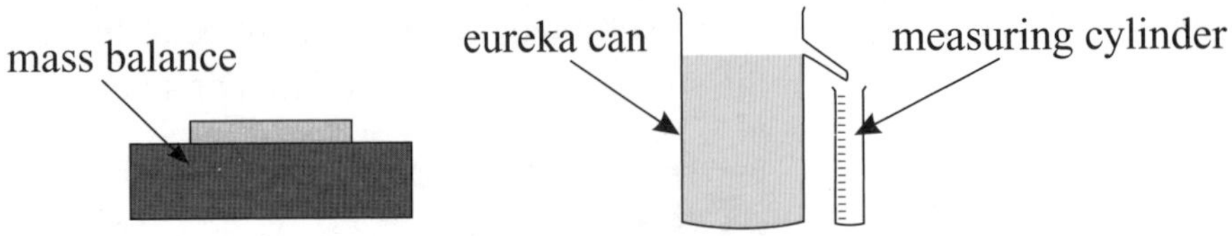

mass balance   eureka can   measuring cylinder

3.1* A eureka can is a beaker with a spout through which overflowing water can escape. The can is filled up to the spout so that when an object is placed in the can the displaced water flows into the measuring cylinder.
Describe the procedure for determining the density of an object using the apparatus in **Figure 1**.

....

....

....

....

....

[4]

3.2 **Table 1** shows an incomplete table of the student's results.

Table 1

Ring	Mass (g)	Water displaced (ml)	Material
A	5.7	0.30	....
B	2.7	0.60	....
C	3.0	0.30	....

Complete **Table 1** using the following information:
Density of gold = 19 g/cm³   Density of silver = 10 g/cm³   Density of titanium = 4.5 g/cm³

[4]

[Total 8 marks]

4 A student investigates the density of an aluminium cola can by submerging it in a measuring cylinder of water. When completely submerged, a full can of unopened cola displaces 337 ml of water. The student then empties the can. She finds that it holds 332 ml of cola and that the mass of the empty can is 13.5 g when it is empty.
Calculate the density of aluminium used to make the can.

Grade 6-7

Density = ... g/cm³

[Total 4 marks]

Internal Energy and Changes of State

1 Use words from the box below to complete the passage.
You can only use a word **once** and you do not need to use all the words. *(Grade 4-6)*

| mass | increases | temperature | volume | decreases |

When a system is heated, the internal energy of the system .... This either

increases the .... of the system or causes a change of state. During a change of

state the temperature and .... of the substance remain constant.

[Total 2 marks]

2 A change of state is a physical change. *(Grade 4-6)*

2.1 State the name of the following changes of state:

Gas to liquid: .... Liquid to gas: ....

[1]

2.2 State what is meant by the term 'physical change'.

....

....

[1]

[Total 2 marks]

3 Heating an object increases its internal energy. *(Grade 6-7)*

3.1 State what is meant by the term 'internal energy'.

....

[1]

3.2 Heating an object can increase its temperature.
State **two** things that the increase in a system's temperature depends on.

1. ....

2. ....

[2]

[Total 3 marks]

4 A student fills a test tube with 30 g of water. He heats the water so that it begins to boil and collects all of the water vapour produced via a tube placed into the bung of the test tube. After the test tube has cooled, he finds that the mass of the water in the test tube is now 20 g. State the mass of the water vapour the student collected. Explain your answer. *(Grade 6-7)*

....

....

....

....

[Total 2 marks]

Topic 3 — Particle Model of Matter

Specific Latent Heat

1 An immersion heater is used to boil 0.50 kg of water in a sealed container. **Grade 6-7**

1.1 Define the term 'specific latent heat'.

....

....
[1]

1.2 The lid is removed when the water begins to boil. The immersion heater transfers
1.13 MJ of energy to evaporate all of the water. Calculate the specific latent
heat of vaporisation of water. Use the equation from the Equations List.

Specific latent heat = ... MJ/kg
[3]

[Total 4 marks]

2 **Figure 1** shows a graph of temperature against time as a substance is heated. **Grade 7-9**

Figure 1

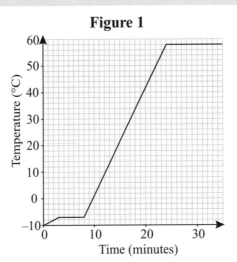

2.1 Describe what is happening during the period 3-8 minutes from the beginning of heating.

....
[1]

2.2 Explain, in terms of particles, why the graph is flat between 3-8 minutes even though the
substance is being heated.

....

....

....
[3]

2.3 Give the melting and boiling points of the substance.

Melting point = ... °C   Boiling point = ... °C   *[2]*

[Total 6 marks]

Particle Motion in Gases

Which of the following statements is true for molecules in a gas? Tick **one** box.

They are constantly moving in all directions at a constant speed. ☐

They are constantly moving in random directions at random speeds. ☐

They are fixed in position. ☐

They all move in the same direction. ☐

1 Use words from the box below to complete the passage. You can use each word **more than** once and you do not need to use all the words.

Grade 4-6

| potential | maximum | decreases | kinetic | increases | average |

When the temperature of a gas increases, the average energy in the .... energy

stores of the gas molecules increases. This .... the .... speed

of the gas molecules. If the gas is kept at a constant volume, increasing the temperature

.... the pressure.

[Total 3 marks]

2 A student investigates how varying the volume of a container full of a fixed mass of gas at a constant temperature affects the pressure of the gas. **Table 1** is an incomplete table of his results.

Grade 6-7

Table 1

Volume (m³)	Pressure (kPa)
8.0×10^{-4}	50
4.0×10^{-4}	100
2.5×10^{-4}	160
1.6×10^{-4}	....

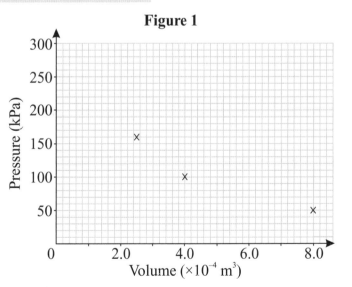

Figure 1

2.1 Complete **Table 1**.

[3]

2.2 Using information from **Table 1**, complete the graph in **Figure 1** by plotting the missing data and drawing a line of best fit.

[2]

[Total 5 marks]

3   **Figure 2** shows a simple piston holding a gas inside a container. The piston is air-tight. The piston is moved and the volume inside the container increases, as shown in **Figure 3**. You can assume that the temperature of the gas inside the container remains constant.

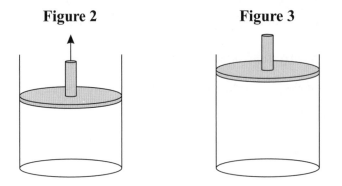

Figure 2   **Figure 3**

Using the particle model, define gas pressure and explain why the gas in the container in **Figure 3** is at a lower pressure than it was in **Figure 2**.

....

....

....

....

....

....

....

[Total 5 marks]

4*   The pistons in a diesel engine work by compressing air and then spraying in droplets of diesel fuel, which then ignites. Explain how compressing the air increases its temperature until it is hot enough to ignite the diesel fuel.

....

....

....

....

....

....

....

....

[Total 6 marks]

Developing the Model of the Atom

Warm-Up

What is the typical radius of an atom?

☐ 1×10^{-10} m   ☐ 1×10^{10} m   ☐ 1×10^{-20} m   ☐ 1×10^{-15} m

How many times smaller is the radius of a nucleus than the radius of the atom?

☐ 10   ☐ 10 000   ☐ 100   ☐ 1000

1   Our understanding of the structure of the atom has   (Grade 4-6)
changed significantly since the early 19th century.

1.1 In 1804, Dalton believed that atoms were tiny spheres which could not be broken up.
State **one** way in which this model is different to our current understanding of atomic structure.

....
[1]

1.2 The alpha particle scattering experiment provided evidence for the nuclear model of the atom.
Name and describe the model that it replaced.

....

....
[2]

1.3 What did the work of James Chadwick prove the existence of around
20 years after the atomic nucleus became an accepted scientific theory?

....
[1]
[Total 4 marks]

2   Niels Bohr discovered that electrons within an   (Grade 4-6)
atom can only exist with defined energy levels.

2.1 Describe how an electron can move between energy levels.

....

....

....
[2]

2.2 Name the type of particle created when an atom loses or gains electrons.

....
[1]

2.3 What is the charge on one of these particles if it is created by an atom losing an electron?

....
[1]
[Total 4 marks]

3* State **two** discoveries about atomic structure which arose from the alpha particle scattering experiment. In each case, state the observation that led to the discovery.

(Grade 6-7)

....

....

....

....

....

....

....

[Total 4 marks]

4 **Table 1** is an incomplete table showing the relative charges of the subatomic particles in an atom.

(Grade 6-7)

Table 1

Particle	Proton	Neutron	Electron
Relative charge	....	....	−1

4.1 Complete **Table 1**.

[2]

4.2 Describe how these subatomic particles are arranged in the atom.

....

....

....

[2]

4.3 An iron atom has 26 protons.
State the number of electrons in the atom and explain your reasoning.

....

....

....

[3]

[Total 7 marks]

Exam Practice Tip

Remember that nothing is ever completely certain — just look at John Dalton and his ideas about atomic structure. New experiments are taking place all the time and they can completely change our models and theories. Make sure you can describe how the nuclear model has changed and how this illustrates the fact that our theories can always change.

Isotopes and Nuclear Radiation

Warm-Up

Draw a line from each form of radiation to show how ionising it is.

gamma   alpha   beta

moderately ionising   weakly ionising   strongly ionising

1   Some isotopes are unstable. They decay into more stable isotopes by emitting nuclear radiation.

Grade 4-6

1.1   What is the name of this process?

....

[1]

1.2   Describe what is meant by isotopes of an element.

....

....

[2]

1.3   Some nuclear radiation is ionising. Define ionisation.

....

[1]

1.4   An unstable isotope releases a particle made up of two protons and two neutrons from its nucleus. Name this type of decay.

....

[1]

[Total 5 marks]

2   An isotope which emits alpha radiation is used in the circuit of a house's smoke detector. Ionising radiation can be damaging if the human body is exposed to it.

Grade 6-7

Explain why the use of ionising radiation in the smoke detector does not pose a threat to the health of people living in the house.

....

....

....

....

[Total 2 marks]

3 One isotope of sodium is $^{23}_{11}\text{Na}$. **Grade 6-7**

3.1 Write down the mass number of this isotope.

....

[1]

3.2 Calculate the number of neutrons in the sodium nucleus.

Number of neutrons = ...

[1]

3.3 Which of the following is another isotope of sodium? Tick **one** box.

$^{11}_{23}\text{Na}$ ☐ $^{11}_{24}\text{Na}$ ☐ $^{23}_{12}\text{Na}$ ☐ $^{24}_{11}\text{Na}$ ☐   *[1]*

3.4 An isotope of neon is $^{23}_{10}\text{Ne}$. Explain whether or not the charge on the neon isotope's nucleus is different to the charge on the nucleus of the sodium isotope.

....

....

....

[2]

[Total 5 marks]

4* Ionising radiation is used to detect leaks in pipes that are buried just below the ground. An unstable isotope is introduced to one end of the pipe and, above the ground, a radiation detector is moved along the path of the pipe. **Grade 7-9**

Explain how this method can be used to identify the location of a leak in the pipe, and suggest what type of radiation the isotope should emit.

....

....

....

....

....

....

....

....

....

....

[Total 6 marks]

Exam Practice Tip

If you're asked about uses of the different kinds of radiation, then think about their properties (how ionising they are, how far they travel etc.). Then just apply what you know to the situation — if you're trying to detect something from a long way away or through a thick barrier, then you want something which has a long range. Simple really...

Nuclear Equations

1 An electron is emitted from a nucleus. **Grade 6-7**

1.1 State the effect this has on the charge of the nucleus.

....

[1]

1.2 Explain the effect that this has on both the mass number and atomic number of the nucleus.

....

....

....

[3]

1.3 After emitting the electron, the atom is excited. It gets rid of excess energy by emitting a gamma ray. What effect does this have on the charge and mass of the nucleus?

....

[1]

[Total 5 marks]

2 A student writes down the following nuclear decay equation: $^{234}_{90}\text{Th} \longrightarrow {}^{234}_{91}\text{Pa} + {}^{0}_{0}\gamma$ **Grade 7-9**

2.1 Explain how you know that this equation is incorrect.

....

[1]

2.2 The student has missed out one other particle which is formed during this decay. Write down the symbol for this particle, including its atomic and mass numbers.

....

[1]

2.3 Radium (Ra) has atomic number 88. The isotope radium-226 undergoes alpha decay to form radon (Rn). Write a nuclear equation to show this decay.

....

[3]

2.4 The radon isotope then undergoes alpha decay to form an isotope of polonium (Po), which undergoes alpha decay to form an isotope of lead (Pb). Calculate the number of neutrons in the nucleus of this lead isotope.

Number of neutrons = ...

[3]

[Total 8 marks]

Half-life

1   The graph in **Figure 1** shows how the count-rate of a radioactive sample changes over time.   Grade 4-6

Figure 1

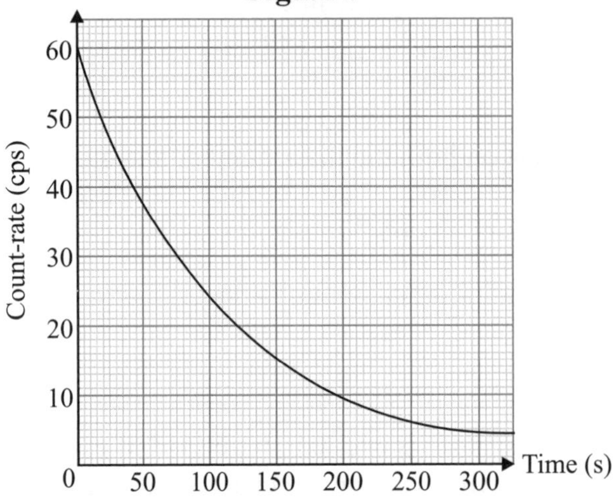

1.1   Define the term 'half-life' in terms of count-rate.

....
[1]

1.2   Using **Figure 1**, determine the half-life of the sample.

Half-life = ... s
[1]

1.3   Initially, the sample contains approximately 800 undecayed nuclei.
      Predict how many of these nuclei will have decayed after two half-lives.

Decayed nuclei = ...
[2]

1.4   After two half-lives, what is the ratio of the number of undecayed
      nuclei left to the initial number of undecayed nuclei?   Tick **one** box.

1:2 ☐   2:1 ☐   1:4 ☐   4:1 ☐   *[1]*

[Total 5 marks]

2   **Table 1** shows data about two radioactive sources.   Grade 6-7

Table 1

	Isotope 1	Isotope 2
Number of undecayed nuclei	20 000	20 000
Half-life	4 minutes	72 years

Explain which isotope will have the highest activity initially.

....
[Total 1 mark]

Topic 4 — Atomic Structure

3 The activity of a radioisotope is 8800 Bq. After 1 hour and 15 minutes, the activity has fallen to 6222 Bq. A further 1 hour and 15 minutes after that, the activity has fallen to 4400 Bq.

3.1 Calculate the radioisotope's half-life. Give your answer in minutes.

Half-life = ... minutes

[1]

3.2 Calculate the activity of the isotope after a total time of 6 hours and 15 minutes has passed. Give your answer to 2 significant figures.

Activity = ... Bq

[2]

[Total 3 marks]

4 A radioactive sample has a 50 second half-life. The initial activity of the sample is 120 Bq.

4.1 Complete the graph in **Figure 2** to show how the activity will change in the first 150 seconds.

Figure 2

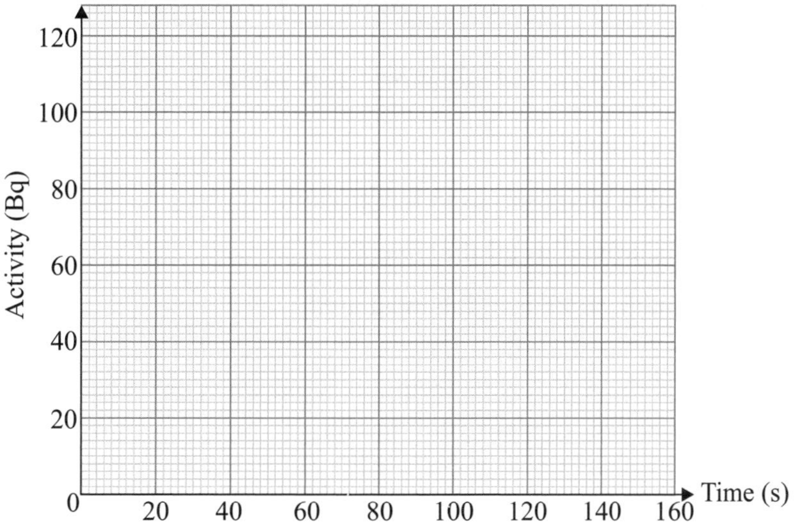

[3]

4.2 Use your graph to predict the activity of the sample after 40 seconds.

Activity = ... Bq

[1]

4.3 Calculate a prediction of the activity after 250 s. Explain why this prediction is less likely to be correct than your prediction in 4.2.

....

....

....

[3]

[Total 7 marks]

Topic 4 — Atomic Structure

Background Radiation and Contamination

Name **one** natural source of background radiation.

....

1   Workers in a nuclear power station take many precautions to prevent unnecessary exposure to radiation. Suggest **two** methods that could be used to reduce their exposure to radiation when dealing with highly radioactive substances.

Grade
4-6

1. ....

2. ....

[Total 2 marks]

2   A physicist is investigating the radioactivity of a sample. She measures the background radiation before the experiment and subtracts it from her measurements.

Grade
4-6

2.1   What is meant by the term background radiation? Tick **one** box.

The levels of alpha radiation nearby. ☐

Radiation produced by radon gas reserves in the ground. ☐

Low-level radiation that is around us all the time. ☐

All radiation from man-made sources in the local area. ☐

[1]

2.2   State the type of error that the physicist is trying to avoid when she subtracts the background radiation level from her results. Explain your answer.

....

....

....

[2]

2.3   State the name given to the amount of radiation that an individual is exposed to.

....

[1]

2.4   State **two** things that can affect the amount of radiation that an individual is exposed to.

1. ....

2. ....

[1]

[Total 5 marks]

3 A scientist is reviewing the safety procedures to be used in her lab. She is concerned about **contamination** and **irradiation**.

_{Grade}
6-7

3.1 Explain the difference between contamination and irradiation.

....

....

....

....

[2]

3.2 Give **one** example of how the scientist can protect herself from being irradiated by a radioactive sample with a low activity.

....

[1]

3.3 Give **two** ways in which the scientist can protect herself against contamination when handling a radioactive sample with a low activity.

1. ....

2. ....

[2]

[Total 5 marks]

4* Radium-226 is an alpha source that was used in clocks until the 1960s to make the hands and numbers glow. Explain whether a clockmaker should be more concerned about irradiation or contamination when repairing old clocks that contain radium.

_{Grade}
7-9

....

....

....

....

....

....

....

....

....

[Total 6 marks]

Exam Practice Tip

It's important to remember that contamination and irradiation aren't the same thing. In the exam, you'll have to make sure you use the correct term when explaining your answers. Go back over your notes if you're unsure, then have another go at the questions on these pages. Then you can reward yourself with a cuppa and a biscuit. Smashing.

😕 ☐   🙂 ☐   😃 ☐

Uses and Risk

Choose some of the words on the left to fill in the blanks on the right.

mutate   die   survive   diagnose   radiation   cancer   sickness

Radiation can cause cells to .... or .... , which can cause cancer or radiation sickness. Radiation can also be used to treat .... and to .... illnesses.

1   Radiotherapy is used often used in the treatment of cancer. Radiation is directed towards the cancerous cells from outside of the body to kill them.

Grade 6-7

1.1   What type of ionising radiation could be used in this procedure? Tick one box.

alpha ☐   beta ☐   gamma ☐   background ☐   *[1]*

1.2   The beam of radiation can be rotated around the patient, keeping the cancerous cells at the centre. Suggest how this method can minimise the risks of radiotherapy.

....

....
[2]

[Total 3 marks]

2   Sources of radiation can be used in medical imaging to explore internal organs. Iodine-123 is a radioactive isotope that is absorbed by the thyroid. Grave's disease causes an overactive thyroid, which causes the thyroid to absorb more iodine than usual.

Grade 7-9

2.1   Briefly explain how iodine-123 could be used to determine if a patient has Grave's disease.

....

....

....

....
[3]

2.2   Iodine-123 emits gamma radiation.
Explain why an alpha emitter would not be used for medical imaging.

....

....
[2]

2.3   Explain why an isotope with a short half-life must be used in this type of procedure.

....

....
[1]

[Total 6 marks]

☹ ☐   😐 ☐   😉 ☐

Fission and Fusion

1 Below are two statements about nuclear fusion. *(Grade 4-6)*

Statement 1: During fusion, a heavier nucleus is formed by joining two lighter nuclei.
Statement 2: During fusion, some mass is converted into energy.

Which of the following is true? Tick **one** box.

Neither statement is true. ☐

Only statement 1 is true. ☐

Only statement 2 is true. ☐

Both statements are true. ☐

[Total 1 mark]

2 State **one** similarity and **one** difference between nuclear fission and nuclear fusion. *(Grade 4-6)*

Similarity: ....

....

Difference: ....

....

[Total 2 marks]

3 Fission reactors use chain reactions to produce energy. *(Grade 6-7)*

3.1* Briefly explain how the absorption of a neutron can lead to a chain reaction.

....

....

....

....

....

....

....

[4]

3.2 Chain reactions have to be controlled. This means limiting the number of neutrons causing fission. Explain what could happen if a chain reaction is uncontrolled.

....

....

[2]

[Total 6 marks]

Contact and Non-Contact Forces

Warm-Up

Write each word below in the table on the right to show whether it is a scalar or vector quantity.

acceleration   time   temperature

mass   weight   force

Scalar	Vector

1 Which of the following correctly defines a vector? Tick **one** box. *(Grade 4-6)*

Vector quantities only have magnitude. ☐

Vector quantities show direction but not magnitude. ☐

Vector quantities have both magnitude and direction. ☐

Vector quantities are a push or pull on an object. ☐

[Total 1 mark]

2 A child is pulling a toy train along the floor by a piece of string. State **one** contact force and **one** non-contact force that acts on the toy. *(Grade 6-7)*

Contact force: ....

Non-contact force: ....

[Total 2 marks]

3 **Figure 1** shows a pair of identical magnets. There is a force of repulsion between them. *(Grade 6-7)*

Figure 1

Magnet A   Magnet B

S   N   N→   S

3.1 Complete the diagram in **Figure 1** by drawing another arrow representing the force that magnet B exerts on magnet A.

[2]

3.2 Magnet B is replaced by a much stronger magnet but magnet A remains the same. Describe how you would redraw the arrows on the diagram to show this new force interaction.

....

....

[2]

[Total 4 marks]

Weight, Mass and Gravity

1   Use words from the box below to complete the passage.
You can only use a word **once** and you do not need to use all the words. *(Grade 4-6)*

| weight | kilograms | mass | directly | inversely | newtons | newton metres |

.... is the amount of matter in an object. .... is

a force due to gravity. Mass is measured in .... whilst weight is

measured in .... The weight of an object is ....

proportional to its mass.

[Total 3 marks]

2   What is meant by the term 'centre of mass'? *(Grade 4-6)*

....

....

[Total 1 mark]

3   The Opportunity rover is a robot which is currently on the surface of the planet Mars. *(Grade 6-7)*
The total mass of the Opportunity rover and its landing parachute is 350 kg.

3.1   Write down the equation that links weight, mass and gravitational field strength.

....

[1]

3.2   Calculate the total weight of the Opportunity rover and its parachute when it was on the Earth.
(The gravitational field strength of the Earth = 9.8 N/kg.)

Weight = ... N

[2]

3.3   When Opportunity landed on Mars it left behind its parachute and moved away to explore.
The mass of the parachute was 209 kg. Calculate the weight of Opportunity without its
parachute on Mars. (The gravitational field strength of Mars = 3.8 N/kg.)
Give your answer to 3 significant figures.

Weight = ... N

[3]

[Total 6 marks]

Exam Practice Tip

Outside of physics, people often use the term weight when they mean mass. Make sure you get the differences straight in your head. You measure mass in on a set of scales, but weight is a force measured by a spring-balance (newtonmeter).

Resultant Forces and Work Done

1   **Figure 1** shows four runners who are running in windy weather.
Tick the box under the runner who is experiencing the largest resultant force.

Grade 4-6

Figure 1

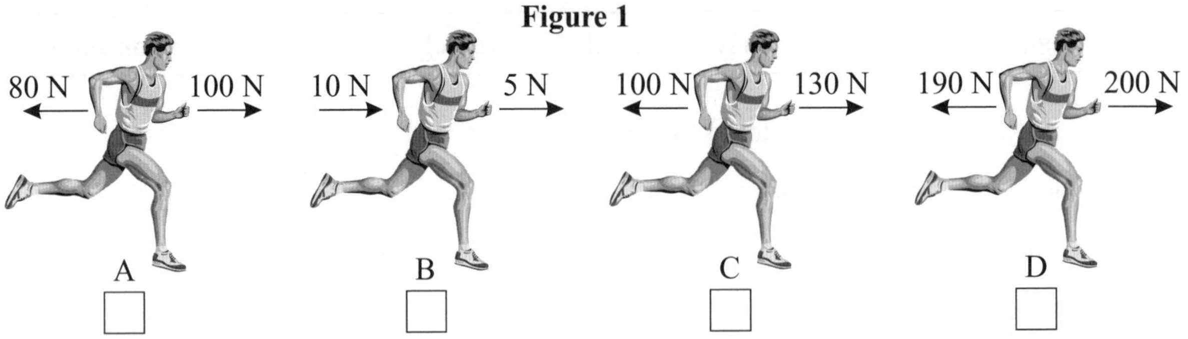

80 N ← → 100 N   10 N → 5 N →   100 N ← → 130 N   190 N ← → 200 N

A ☐   B ☐   C ☐   D ☐

[Total 1 mark]

2   A woman pulls a 20 kg suitcase along a 15 m corridor using a horizontal force of 50 N.   **Grade 6-7**

2.1   Calculate the work done by the woman. Give the correct unit.

Work done = ... Unit = ...
[3]

2.2   Work has to be done against frictional forces acting on the wheels of the suitcase.
Explain the effect this has on the temperature of the suitcase.

....

....
[2]

[Total 5 marks]

3   **Figure 2** shows an incomplete free body diagram of a ladder leaning
against a wall. There is no friction between the ladder and the
wall but there is friction between the ladder and the ground.   **Grade 7-9**

Figure 2

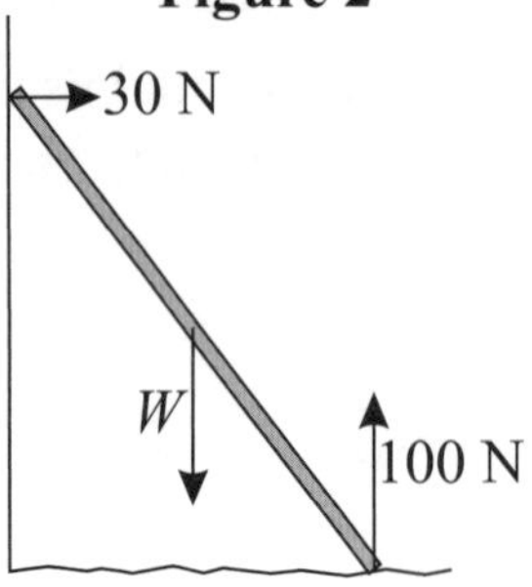

→ 30 N

W ↓   ↑ 100 N

3.1   Using **Figure 2**, determine the weight of the ladder, W.

Weight = ... N
[1]

3.2   Complete **Figure 2** by drawing the missing frictional force.

[2]

[Total 3 marks]

:(☐   :| ☐   :) ☐

Calculating Forces

Find the horizontal and vertical components of the force shown on the right. Each side of a square equals 1 N.

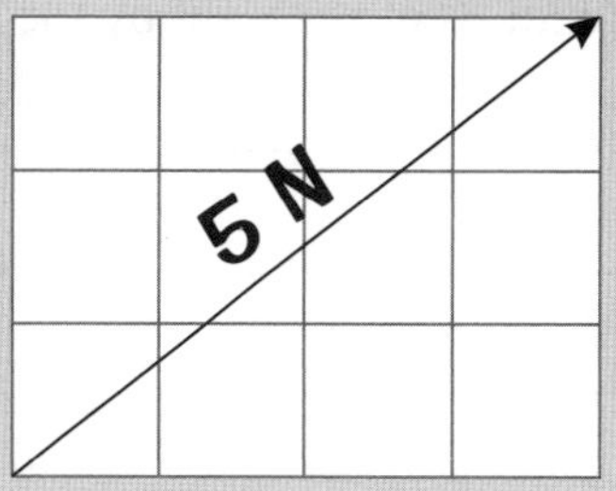

Horizontal component = ... N

Vertical component = ... N

1   **Figure 1** shows a girl on a swing. Her weight of 500 N acts vertically downwards and a tension force of 250 N acts on the ropes at an angle of 30° to the horizontal.

Figure 1

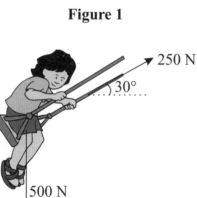

250 N

30°

500 N

Figure 2

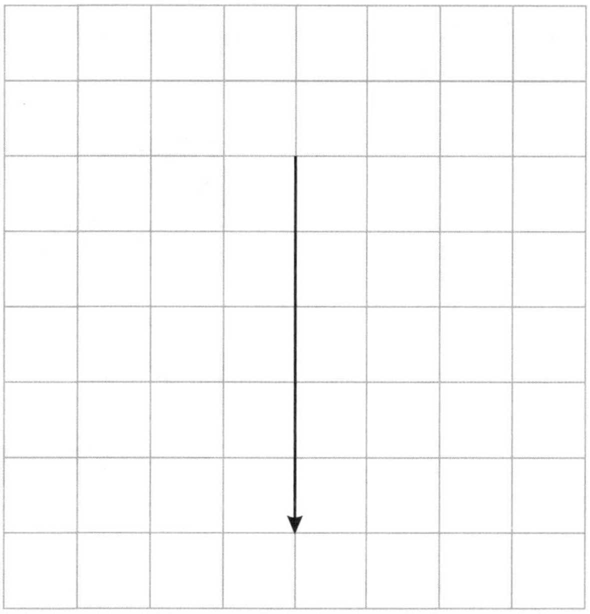

1.1 **Figure 2** shows an incomplete scale drawing for the forces acting on the girl.
Only the girl's weight has been drawn so far. Calculate the scale used in the drawing.

.... cm = ... N
[1]

1.2 Complete the scale drawing in **Figure 2** to find the magnitude of the resultant force acting on the girl.

Magnitude = ... N
[2]

[Total 3 marks]

Forces and Elasticity

1   Deformations can be elastic or inelastic. Grade 4-6

1.1   Explain what is meant by the terms elastic deformation and inelastic deformation.

....

....

....

....
[2]

1.2   Stretching is one way in which forces can deform an object. State **two** other ways.

....
[1]

[Total 3 marks]

2   A student investigates the change in height of a toy horse in a playground, shown in **Figure 1**, when different people sit on it. Grade 7-9

Figure 1

2.1   When a child weighing 250 N sits on the toy horse in **Figure 1**, his feet don't touch the floor.
The height of the toy horse decreases by 20 cm. Calculate the spring constant of the spring.
Give the correct unit.

Spring constant = ... Unit = ...
[3]

2.2   The child gets off and the student's teacher then sits on the toy horse. Her weight is double
that of the child. The student predicts that the height of the toy horse will change by 40 cm.
Explain whether or not you agree with the student. State any assumptions you have made.

....

....

....
[2]

[Total 5 marks]

Investigating Springs

PRACTICAL

1 A student carried out an investigation to study the relationship between the force exerted on and the extension of a spring. He hung different numbers of 1 N weights from the bottom of the spring and measured the extension of the spring with a ruler, as shown in **Figure 1**. *(Grade 6-7)*

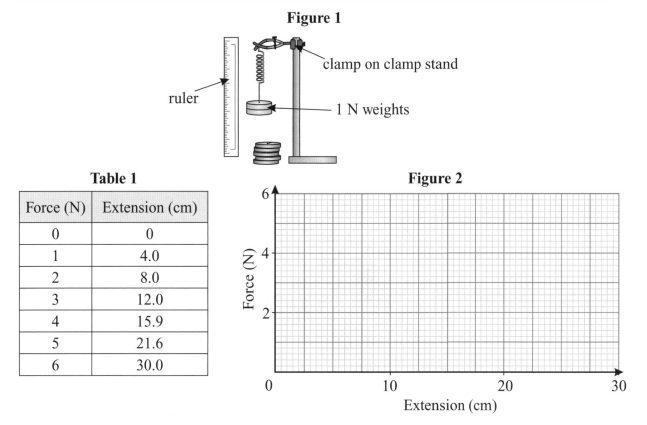

Figure 1

ruler — clamp on clamp stand — 1 N weights

Table 1

Force (N)	Extension (cm)
0	0
1	4.0
2	8.0
3	12.0
4	15.9
5	21.6
6	30.0

Figure 2

1.1 **Table 1** shows the results that the student obtained in his investigation.
Draw the force-extension graph for the student's results on the axes in **Figure 2**.

[3]

1.2 Using the graph you have drawn, calculate the spring constant of the spring being tested.

Spring constant = ... N/m
[2]
[Total 5 marks]

2 Calculate the work done on a spring when it is extended elastically by 8.0 cm. The spring constant of the spring is 25 N/m. *(Grade 6-7)*

Work done = ... J
[Total 2 marks]

Exam Practice Tip

You need to know this practical really well — you could be asked pretty much anything about it in the exam. And make sure you draw graphs accurately with a sharp pencil. It'll really help if you need to use the graph to work something out.

Moments

1 **Figure 1** shows a box spanner used by a mechanic. He applies a force of 50 N at the end of the spanner. Calculate the size of the moment created, stating any equations you use.

Grade 4-6

Figure 1

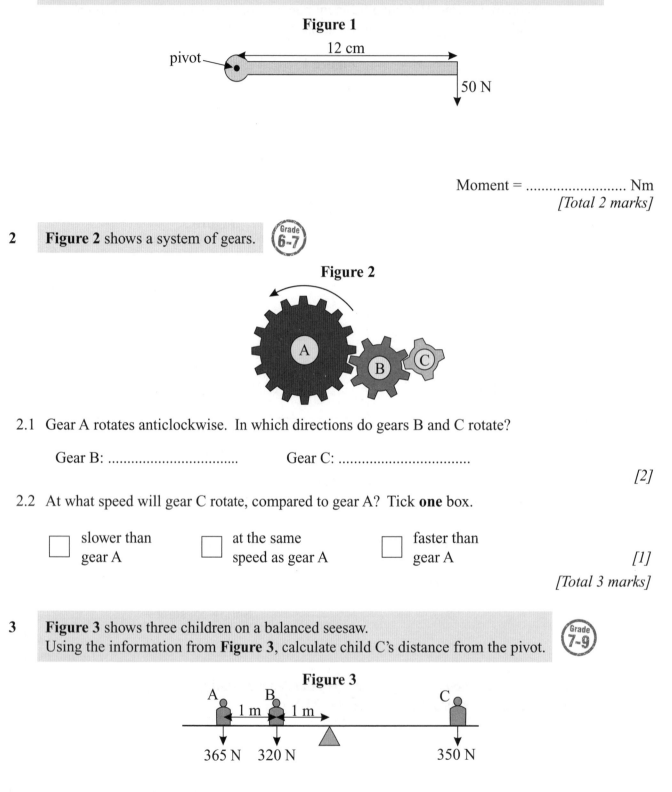

Moment = ... Nm

[Total 2 marks]

2 **Figure 2** shows a system of gears.

Grade 6-7

Figure 2

2.1 Gear A rotates anticlockwise. In which directions do gears B and C rotate?

Gear B: .... Gear C: ....

[2]

2.2 At what speed will gear C rotate, compared to gear A? Tick **one** box.

☐ slower than gear A ☐ at the same speed as gear A ☐ faster than gear A

[1]

[Total 3 marks]

3 **Figure 3** shows three children on a balanced seesaw.
Using the information from **Figure 3**, calculate child C's distance from the pivot.

Grade 7-9

Figure 3

A   B   C
1 m   1 m

365 N   320 N   350 N

Distance = ... m

[Total 3 marks]

Fluid Pressure

Choose from the labels on the left to fill in the blanks on the right.
You do not need to use all of the words.

parallel   gases   volume

solids   perpendicular

parsecs   area   pascals

Liquids and .... are both fluids.

Fluid pressure is the force exerted ....

to a surface, per unit ....

The unit of pressure is ....

---

1 Water in the chamber of a water pistol is squeezed out when a force is applied by the plunger. A force of 12 N is applied by a plunger with a surface area of 0.15 m². *(Grade 4-6)*

1.1 Write down the equation that links pressure, force and area.

....

[1]

1.2 Calculate the pressure at the plunger.

Pressure = ... Pa

[2]

[Total 3 marks]

---

2 **Figure 1** shows a spouting can. A student fills the can with water and allows it to drain. Explain why water escaping from the bottom of the spouting can does so at a faster rate than water from the top. *(Grade 6-7)*

Figure 1

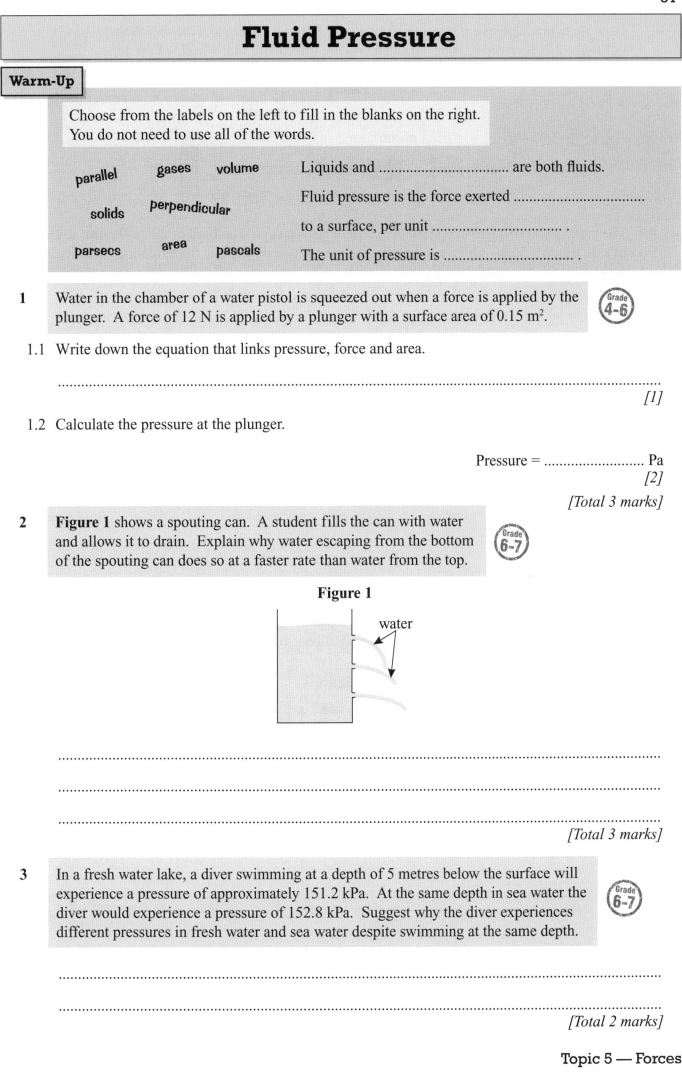

water

....

....

....

[Total 3 marks]

---

3 In a fresh water lake, a diver swimming at a depth of 5 metres below the surface will experience a pressure of approximately 151.2 kPa. At the same depth in sea water the diver would experience a pressure of 152.8 kPa. Suggest why the diver experiences different pressures in fresh water and sea water despite swimming at the same depth. *(Grade 6-7)*

....

....

[Total 2 marks]

4   Cars use hydraulic braking systems. A hydraulic braking system like the one shown
     in **Figure 2** is designed to apply an equal braking force to all four wheels of the car.

Figure 2

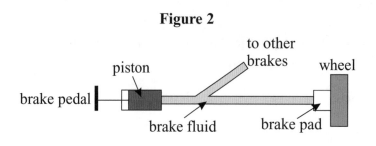

4.1 The brake pedal of a four-wheel car is connected to a piston. The head of the piston is a square,
     of side length 2.5 cm. Calculate the pressure transmitted throughout the hydraulic fluid when the
     brake is applied with a force of 100 N.

Pressure = ... Pa

[3]

4.2 The brake pad attached to one wheel has a surface area of 0.005 m².
     Calculate the total braking force acting on the car.

Force = ... N

[3]

[Total 6 marks]

5   The bottle of lemonade shown in **Figure 3** is full to the brim.
     Calculate the change in pressure in the bottle between point X and
     the base of the bottle. Give your answer to 2 significant figures.
     Lemonade has a density of 1000 kg/m³. Gravitational field strength = 9.8 N/kg.

Figure 3

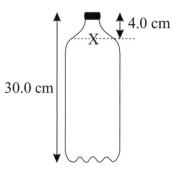

Change in pressure = ... Pa

[Total 3 marks]

Upthrust and Atmospheric Pressure

1   Use words from the box below to complete the passage.   Grade 4-6
    You can only use a word **once** and you do not need to use all the words.

decreases	thrust	larger	increases	upthrust	smaller

In a liquid, pressure .... with depth. This means that the force acting on the

bottom of a submerged object is .... than the force acting on the top of the

object. This leads to a resultant force called ....

[Total 2 marks]

2   **Figure 1** shows a digital force gauge being used to measure   Grade 6-7
    the weight of a golf ball in air and whilst it is places in water.

Figure 1

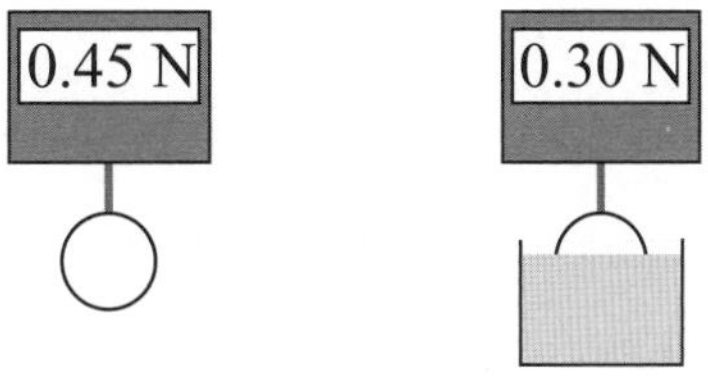

Explain why the weight of the golf ball appears to change when it is placed in the water.

....

....

[Total 3 marks]

3   A silver necklace is dropped into the ocean. Explain why the necklace sinks.   Grade 6-7
    The density of silver is 10 490 kg/m^3. The density of the water is 1000 kg/m^3.

....

....

....

[Total 3 marks]

4* Explain why atmospheric pressure decreases with altitude.   Grade 7-9

....

....

....

....

....

....

....

[Total 6 marks]

Topic 5 — Forces

Distance, Displacement, Speed and Velocity

Choose from the words on the left to fill in the blanks on the right. Use each word once.

distance

velocity

vector

scalar

Displacement and .... are both

.... quantities. This means they have both a

size and a direction. Speed and .... are both

.... quantities. They do not depend on direction.

1   **Figure 1** shows the path taken by a football kicked by a child. When it is kicked at Point A, the ball moves horizontally to the right until it hits a vertical wall at Point B. The ball then bounces back horizontally to the left and comes to rest at Point C.

Grade 4-6

Figure 1

Scale 1 cm = 1 m

A   C   B

1.1 What is the distance that the ball has moved through from A to B?

Distance = ... m

[1]

1.2 What is the total distance that the ball has moved through from A to C?

Distance = ... m

[1]

1.3 Draw a vector arrow on **Figure 1** to show the displacement of the ball.

[1]

1.4 What is the magnitude of the displacement of the ball after it has come to rest?

Displacement = ... m

[1]

[Total 4 marks]

2   The speed of sound varies depending upon the substance it is travelling through. State the speed of sound in air.

Grade 4-6

....

[Total 1 mark]

3   Give **three** factors that can affect a person's walking, running or cycling speed.

Grade 4-6

....

....

[Total 3 marks]

4 Explain whether a satellite orbiting the Earth at 3.07×10^3 m/s has a constant velocity.

(Grade 6-7)

....

....

[Total 2 marks]

5 A man has just got a new job and is deciding whether to walk, cycle or take a bus to get to work. There are two routes he could take. The shorter route is along a 6 km path that only pedestrians and cyclists are allowed to use. The bus takes a longer route along a road.

(Grade 6-7)

5.1 Write down the formula that links distance travelled, speed and time.

....
[1]

5.2 Estimate how long it would take the man to walk the pedestrian route.

Time taken = ... s
[3]

5.3 Estimate how much time would be saved if the man cycled this route instead.

Time saved = ... s
[4]

5.4 Travelling to work by bus takes 20 minutes.
The total distance covered during this time is 9.6 km.
Calculate the average speed of the bus.

Average speed = ... m/s
[3]
[Total 11 marks]

6 The speed at which an aircraft flies is often expressed in terms of its Mach number, which describes the speed in relation to the speed of sound. For example, Mach 2 is twice the speed of sound. A commercial airliner on a long-haul flight has a speed of Mach 0.8. The temperature of the air is typically –60 °C.

(Grade 7-9)

The speed of sound is temperature dependent and can be found using:

Speed of sound in m/s = $331 + 0.6T$, where T is the temperature in °C.

Calculate the distance travelled by the jet over 5.0×10^4 s.

Distance travelled = ... km
[Total 4 marks]

Topic 5 — Forces

Acceleration

Draw one line from each scenario to the typical acceleration for that object.

A sprinter starting a race	10 m/s²
A falling object	2 × 10⁵ m/s²
A bullet shot from a gun	1.5 m/s²

1 Briefly describe the motion of a decelerating object. *Grade 4-6*

....

[Total 1 mark]

2 **Table 1** shows how the speed of a car changes with time as it accelerates uniformly. *Grade 6-7*

Table 1

Time (s)	0	1	2	3
Speed (m/s)	0	4	8	12

2.1 Write down the formula that links acceleration, velocity and time.

....

[1]

2.2 Calculate the acceleration of the car.

Acceleration = ... m/s²

[2]

[Total 3 marks]

3 A car accelerates uniformly at 2.5 m/s² from rest to a speed of 20 m/s. Calculate the time taken for the car to reach 20 m/s. *Grade 6-7*

Time = ... s

[Total 3 marks]

4 A train travelling at 32 m/s slows down to 18 m/s over a distance of 365 m. Calculate the deceleration of the train over this distance. Use an equation from the Equations List. *Grade 7-9*

Deceleration = ... m/s²

[Total 2 marks]

Distance-Time and Velocity-Time Graphs

1   A boat is being rowed along a straight canal. Some students use a watch to time how long after setting off the boat passes marker posts spaced 100 metres apart. **Table 1** shows their results.

Table 1

Distance (m)	0	100	200	300	400	500
Time (s)	0	85	165	250	335	420

Figure 1

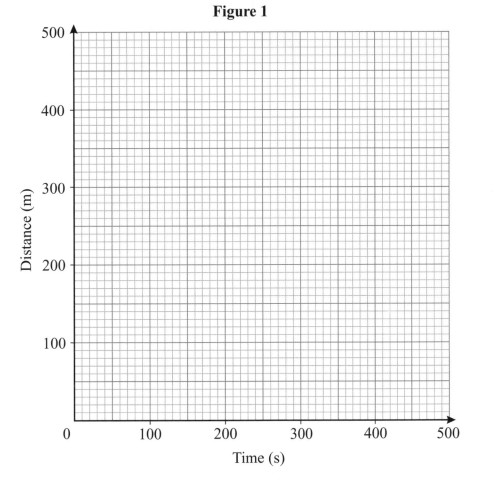

1.1 Draw the distance-time graph for the results in **Table 1** on the axes shown in **Figure 1**.

[3]

1.2 Using the graph in **Figure 1**, determine how far the boat travelled in 300 s.

Distance = ... m

[1]

1.3 Determine how long it took the boat to travel 250 m.

Time = ... s

[1]

1.4 Suggest **one** way to make the timings made by the students more accurate.

....

....

[1]

[Total 6 marks]

Topic 5 — Forces

2   **Figure 2** shows the distance-time graph for a cyclist's bike ride.   (Grade 6-7)

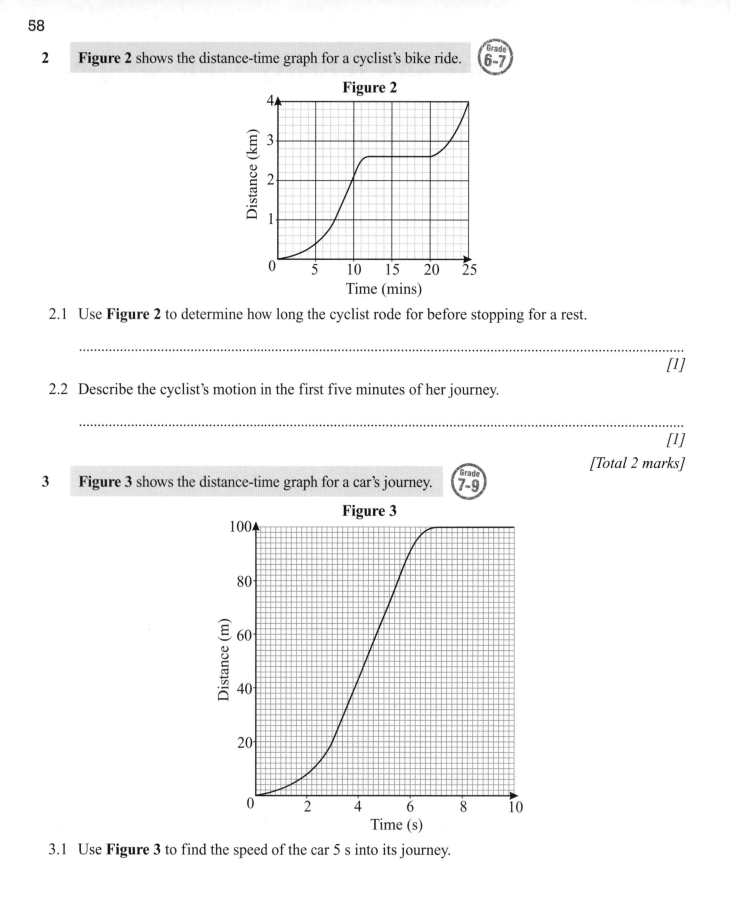

Figure 2

2.1 Use **Figure 2** to determine how long the cyclist rode for before stopping for a rest.

....
[1]

2.2 Describe the cyclist's motion in the first five minutes of her journey.

....
[1]

[Total 2 marks]

3   **Figure 3** shows the distance-time graph for a car's journey.   (Grade 7-9)

Figure 3

3.1 Use **Figure 3** to find the speed of the car 5 s into its journey.

Speed = ... m/s
[3]

3.2 Use **Figure 3** to find the speed of the car 2 s into its journey.

Speed = ... m/s
[3]

[Total 6 marks]

4   **Figure 4** shows an incomplete velocity-time graph for a roller coaster ride.   (Grade 7-9)

Figure 4

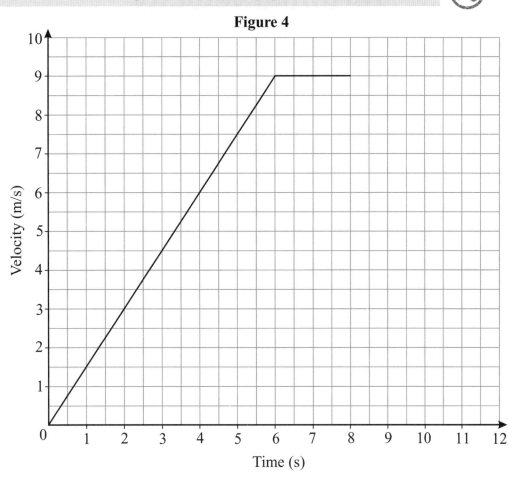

4.1 After 8 seconds, the roller coaster decelerates at an increasing rate.
It comes to rest 4 seconds after it begins decelerating.
Complete the velocity-time graph in **Figure 4** to show this.

[2]

4.2 Calculate the acceleration of the roller coaster during the first 6 seconds of the ride.

Acceleration = ... m/s^2
[2]

4.3 Calculate the distance travelled by the ride between 0 and 8 s.

Distance = ... m
[4]

4.4 Calculate the distance travelled during the entire ride to the nearest metre.

Distance = ... m
[5]
[Total 13 marks]

Topic 5 — Forces

Terminal Velocity

1 Any object falling (in a fluid) for long enough reaches its terminal velocity. Which statements correctly describe terminal velocity? Tick **two** boxes.

Grade 4-6

Terminal velocity is the minimum velocity an object can fall at. ☐

The resultant vertical force on an object falling at its terminal velocity is zero. ☐

The resultant vertical force on an object falling at its terminal velocity equals its weight. ☐

Terminal velocity is the maximum velocity an object can fall at. ☐

[Total 1 mark]

2 A ball is dropped and falls for 6 seconds before reaching its terminal velocity of 40 m/s. After 15 seconds the ball hits the ground. Draw a velocity-time graph for the first ten seconds of the ball's motion on the axes shown in **Figure 1**.

Grade 6-7

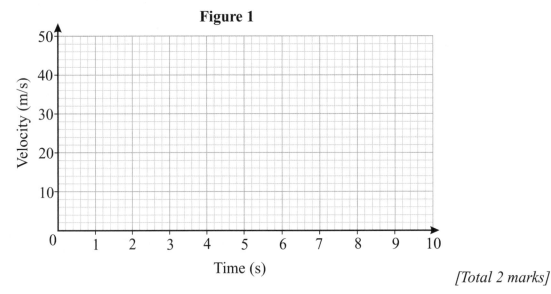

Figure 1

[Total 2 marks]

3 A student drops a large book and a cricket ball that both have the same weight from a tall building. Explain why both objects eventually fall at a constant velocity and why the terminal velocity of the book is lower than the terminal velocity of the ball.

Grade 7-9

....

....

....

....

....

....

[Total 5 marks]

Exam Practice Tip

Remember that the acceleration of a falling object is continuously decreasing due to air resistance until it reaches zero. That means you won't be able to use any of those fancy equations for uniform acceleration that you're used to.

☹ ☐   😐 ☐   🙂 ☐

Newton's First and Second Laws

1 State Newton's First Law for a stationary object. (Grade 4-6)

....

....

[Total 1 mark]

2 Use words from the box below to complete the passage. You can only use a word **once** and you do not need to use all of the words. (Grade 4-6)

| area | mass | inversely | directly | resistive | resultant |

Newton's Second Law states that the acceleration of an object is ....

proportional to the .... force acting on the object and

.... proportional to the .... of the object.

[Total 3 marks]

3 **Figure 1** shows the horizontal forces acting on a motorbike travelling at a constant velocity. (Grade 6-7)

Figure 1

3.1 There are two resistive forces acting on the bike. Suggest what these forces may be.

....

....

[2]

3.2 The engine provides a driving force of 5.0 kN. One of the resistive forces has a magnitude of 3.85 kN. Calculate the size of the second resistive force.

Force = ... N

[1]

[Total 3 marks]

4 A 5.0 kg vase is knocked from a shelf. **Grade 6-7**

4.1 Write down the formula that links force, mass and acceleration.

....

[1]

4.2 Calculate the resultant force acting on the vase as it begins to fall.
Acceleration due to gravity, $g = 9.8$ m/s^2.

Force = ... N

[2]

[Total 3 marks]

5 A 1450 kg car accelerates uniformly from rest. It reaches 24 m/s in 9.2 s.
Calculate the force needed to cause this acceleration. **Grade 6-7**

Force = ... N

[Total 4 marks]

6 **Figure 2** shows a 7520 kg lorry. The driver spots a hazard ahead and applies
the brakes. The lorry decelerates uniformly and comes to a stop 50 m after
the brakes are applied. Estimate the braking force needed to stop the lorry. **Grade 7-9**

Figure 2

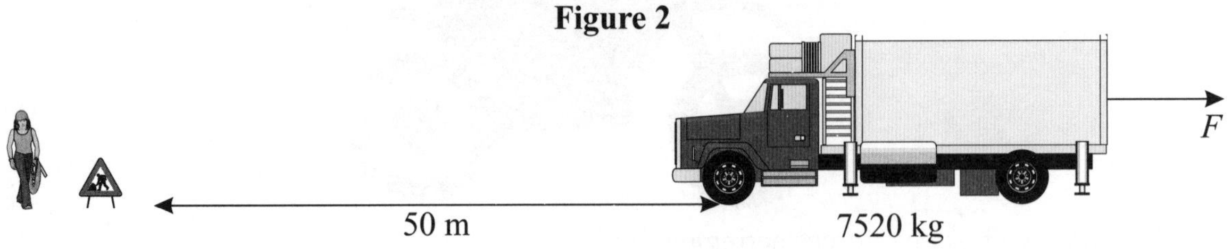

50 m   7520 kg F

Force = ... N

[Total 5 marks]

Exam Practice Tip

Watch out for questions talking about constant or uniform acceleration over a distance. They can be tricky and require a
lot of steps. If you're struggling, read the question carefully, pick out the key bits of information and write them all down.
Then look on the equation sheet to see if there are any equations you can use to find the values the question is asking for.

Inertia and Newton's Third Law

Which of the following is Newton's Third Law? Tick **one** box.

A non-zero resultant force is needed to cause a change in speed or direction. ☐

A resultant force is inversely proportional to the mass of an object. ☐

When two objects interact, they exert equal and opposite forces on each other. ☐

A resultant force of zero leads to an equilibrium situation. ☐

1 **Figure 1** shows the forces acting on a gymnast in equilibrium balancing on two beams. *Grade 4-6*

Figure 1

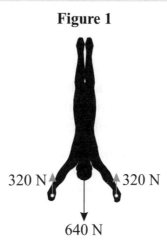

320 N   320 N

640 N

1.1 State the force exerted by each of the gymnast's hands on the balance beams.

Force = ... N

[1]

1.2 State the name of this force.

....

[1]

1.3 State the size of the attractive force exerted on the Earth by the gymnast.

Force = ... N

[1]

[Total 3 marks]

2 Define the following terms: *Grade 6-7*

2.1 Inertia

....

[1]

2.2 Inertial mass

....

[1]

[Total 2 marks]

Investigating Motion

1   **Figure 1** shows the apparatus used by a student to investigate the effect of varying force on the acceleration of a trolley. The trolley is on a frictionless, flat surface.

Grade
6-7

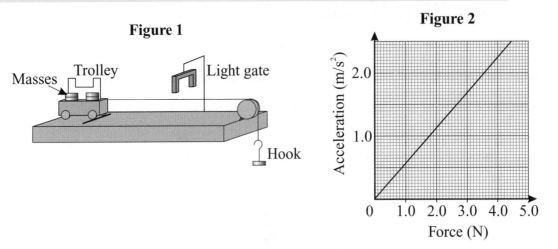

Figure 1

Figure 2

When the hook is allowed to fall, the trolley accelerates. The force acting on, and the acceleration of, the trolley are recorded. The student changes the force on the trolley by moving a mass from the trolley to the hook. **Figure 2** is a graph of acceleration against force for the trolley.

1.1 Give **one** conclusion that can be made from **Figure 2**.

....
[1]

1.2 Write down the formula that links force, mass and acceleration.

....
[1]

1.3 Calculate the mass of the system from **Figure 2**.

Mass = ... kg
[3]

[Total 5 marks]

2   A second student investigates how the mass of a trolley affects its motion down a fixed ramp. The accelerating force on the trolley is the component of the trolley's weight that acts along the ramp. The student adds masses to the trolley and each time measures its final speed at the bottom of the ramp to calculate its acceleration. Explain why this experiment will not correctly show the relationship between the mass and acceleration of the trolley.

Grade
7-9

....

....

....

....

....

[Total 2 marks]

Stopping Distances

1 Define the following terms: (Grade 4-6)

1.1 Thinking distance

....

[1]

1.2 Braking distance

....

[1]

[Total 2 marks]

2 The thinking distance for a driver in a car travelling at 40 mph is 12 m. The braking distance is 24 m. Calculate the car's stopping distance when it is travelling at 40 mph. (Grade 4-6)

Stopping Distance = ... m

[Total 1 mark]

3 When a vehicle's brakes are applied, energy is transferred away from the kinetic energy stores of the wheels. State what causes this and describe the effect it has on the brakes. (Grade 6-7)

....

....

....

....

[Total 2 marks]

4* Explain the importance of a car having brakes and tyres that are in good condition and the effect this will have on stopping distance and safety. (Grade 6-7)

....

....

....

....

....

....

....

....

....

[Total 6 marks]

Topic 5 — Forces

Reaction Times

1 What is the typical reaction time for a person? (Grade 4-6)

☐ 1.3 – 1.8 s   ☐ 0.2 – 0.9 s   ☐ 0.01 – 0.02 s   ☐ 2.0 – 3.0 s

[Total 1 mark]

2 Give **three** things that could affect a person's reaction time. (Grade 4-6)

1. ....

2. ....

3. ....

[Total 3 marks]

3 A teacher tests the reaction times of two of her students by measuring how far a ruler falls vertically before the student catches it. (Grade 4-6)

3.1 Describe **one** other method that can be used to test people's reaction times.

....

[1]

3.2 **Table 1** shows the results. The values in the table show the distance the ruler falls in cm during each attempt. Complete the table by working out the average distance fallen by the ruler for each student.

Table 1

	Attempt 1	Attempt 2	Attempt 3	Average
Student A	7.0	7.1	6.9	....
Student B	8.4	8.2	8.3	....

[2]

3.3 Which student has the fastest average reaction time? Give a reason for your answer.

....

[1]

3.4 Suggest **two** ways the teacher could make the experiment a fair test.

....

....

[2]

3.5 The teacher then repeats the experiment. This time, she has a third student talk to the student being tested. Predict how this will affect the reaction times of both students A and B.

....

[1]

[Total 7 marks]

4 Describe the steps involved when using the ruler drop experiment to investigate reaction times.

(Grade 6-7)

....

....

....

....

....

....

....

[Total 5 marks]

5* A man is driving home late at night. He listens to loud music as he drives to keep himself alert. He is impatient to get home so drives quickly. Explain the safety implications of the man's actions.

(Grade 6-7)

....

....

....

....

....

....

....

[Total 4 marks]

6 A student tests his reaction time with a metre ruler using a ruler drop experiment. He catches the metre ruler after it has fallen 45.0 cm. Calculate his reaction time.

(Grade 7-9)

The acceleration due to gravity is 9.8 m/s².

Reaction time = ... s

[Total 4 marks]

Exam Practice Tip

For long explanation answers make sure you cover every point mentioned in the question. You won't get all of the marks if you miss out part of what they're asking for, no matter how much you write for the rest of your points.

Topic 5 — Forces

More on Stopping Distances

1   **Table 1** shows how the stopping distance of a car varies with speed. Grade 6-7

Table 1

Speed (mph)	0	10	20	30	40	50	60	70
Stopping distance (m)	0	5	12	23	36	53	73	96

Figure 1

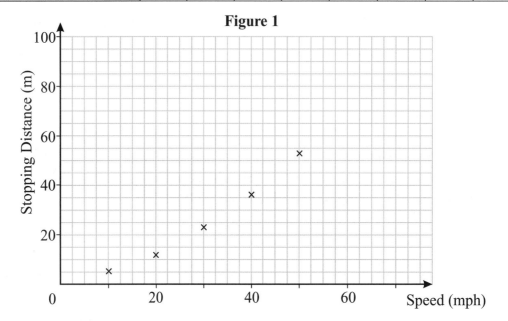

1.1   Using the data in **Table 1**, complete the graph shown in Figure 1, including a line of best fit.

[3]

1.2   Using **Figure 1**, find the stopping distance for a car travelling at 35 mph.

Stopping distance = ... m
[1]

1.3   Calculate how much further the car would travel before stopping if it was travelling at 65 mph.

Distance = ... m
[2]

[Total 6 marks]

2   The stopping distance for a truck travelling at 18 m/s is 45 m. The truck driver has a reaction time of 0.50 s. Estimate the stopping distance for the truck if it were to travel at 36 m/s. Grade 7-9

Stopping distance = ... m
[Total 5 marks]

Momentum

Warm-Up

The snippets below show the parts of a description of momentum.
Number each snippet 1 to 5 to show the correct order. The first one has been done for you.

[ ] ...vector quantity and is equal to...

[1] Momentum is a property of...

[ ] ...moving objects.

[ ] ...mass × velocity.

[ ] It is a...

1 A motorbike is travelling at 25 m/s and has 5500 kg m/s of momentum. *(Grade 4-6)*

1.1 Write down the equation that links momentum, mass and velocity.

....

[1]

1.2 Calculate the mass of the motorbike.

Mass = ... kg

[3]

[Total 4 marks]

2 **Figure 1** and **Figure 2** show a Newton's cradle. *(Grade 6-7)*
All of the balls on the cradle have the same mass.

Figure 1

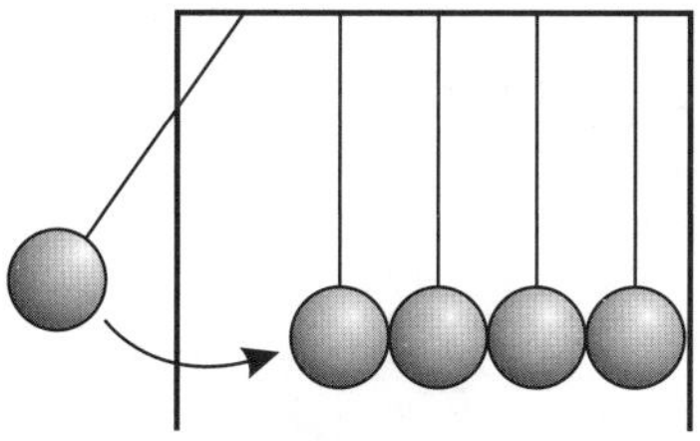

Figure 2

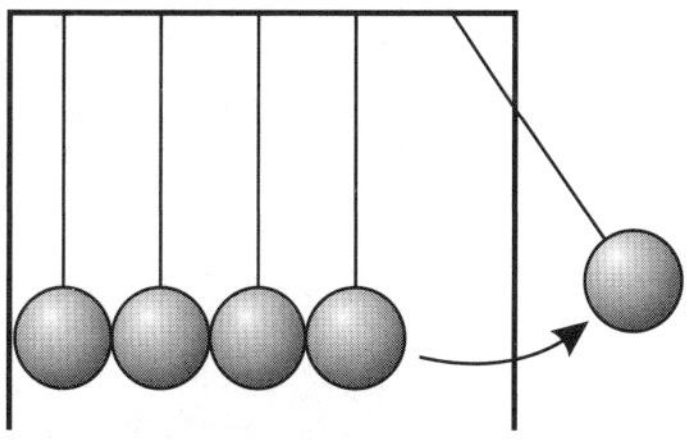

When a ball is lifted and allowed to hit the others as shown in **Figure 1**, it causes the last ball in the line to move outwards, as shown in **Figure 2**. The balls in between appear to remain stationary. The velocity of the first ball when it hits the second ball is equal to the velocity of the final ball when it starts to move. Using conservation of momentum, explain this behaviour.

....

....

....

....

....

....

[Total 4 marks]

Topic 5 — Forces

Changes in Momentum

1 State what the rate of change of an object's momentum is equal to. *(Grade 4-6)*

....

[Total 1 mark]

2 A ball has its momentum changed by 10 kg m/s in 0.1 s. Calculate the force acting on the ball. *(Grade 6-7)*

Force = ... N

[Total 2 marks]

3 During a collision, an air bag is activated in a car. Explain how the air bag reduces the risk of the driver being injured during the collision. *(Grade 6-7)*

....

....

....

....

[Total 4 marks]

4 **Figure 1** shows two American football players running towards each other. They collide and cling together in the tackle. Calculate the velocity that they move together with after the tackle. *(Grade 7-9)*

Figure 1

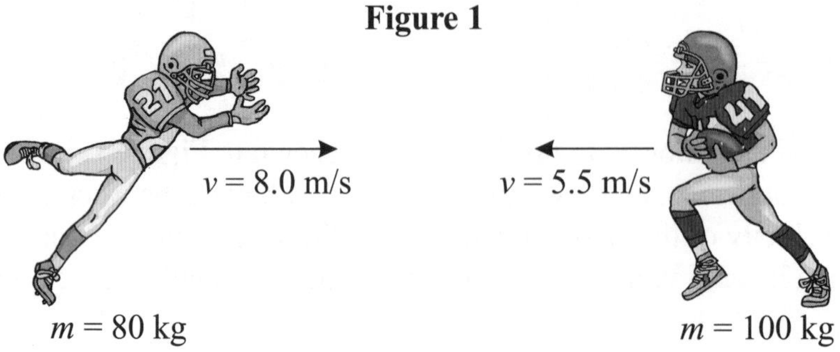

$v = 8.0$ m/s $v = 5.5$ m/s

$m = 80$ kg $m = 100$ kg

Magnitude of velocity = ... m/s

Direction = ...

[Total 5 marks]

Transverse and Longitudinal Waves

1   A student produces two types of waves on a spring, A and B, as shown in **Figure 1**.   (Grade 4-6)

Figure 1

A   B

1.1   State whether each spring shows a transverse or a longitudinal wave.

Wave A: ....   Wave B: ....
[1]

1.2   Label the wavelength of wave A on **Figure 1**.
[1]

1.3   Define the term 'amplitude'.

....
[1]

1.4   Give **one** example of a transverse wave.

....
[1]

[Total 4 marks]

2   **Figure 2** shows a loudspeaker. It produces a sound wave with a frequency of 200 Hz.   (Grade 6-7)

Figure 2

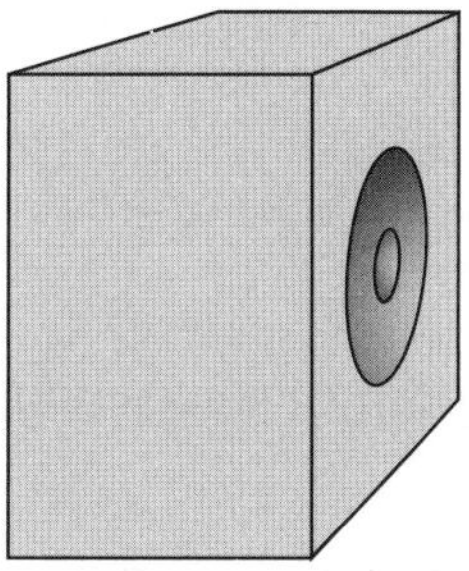

2.1   Draw an arrow on **Figure 2** to show the direction in which the sound wave transfers energy.
[1]

2.2   Calculate the period of the sound wave.

Period = ... s
[2]

2.3   Describe the difference between a longitudinal wave and a transverse wave.

....

....

....
[2]

[Total 5 marks]

Experiments with Waves

1 **Figure 1** shows ripples on the surface of some water in a ripple tank. The signal generator producing the ripples is set to a frequency of 12 Hz. A student measures the distance between the first and last visible ripple as 18 cm, as shown in **Figure 1**.

Grade 6-7

Figure 1

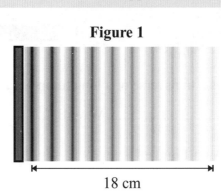

18 cm

1.1 The student finds it difficult to measure the distance because the ripples are moving. Suggest and explain what the student could do to make the measurement easier.

....

....

....

[2]

1.2 Calculate the speed of the ripples in the water.

Speed = ... m/s

[3]

[Total 5 marks]

2* Describe a method to measure the speed of waves on a string. *Grade 7-9*

....

....

....

....

....

....

....

....

[Total 6 marks]

Reflection

At the boundary with a new material, a wave can be reflected, absorbed or transmitted.
Draw a line to match each option to the correct definition.

wave is reflected   it passes through the material

wave is absorbed   it bounces back off the material

wave is transmitted   it transfers all energy to the material

1 **Figure 1** shows two mirrors that meet at 90°. A ray of light hits one of the mirrors.
Complete the diagram to show the path of the light ray as it reflects off both mirrors.
You should draw normal lines to help you construct your diagram.

Grade 6-7

Figure 1

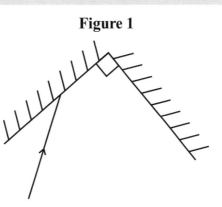

[Total 2 marks]

2 A man is wearing mirrored sunglasses. Anyone looking at him can see their reflection
in the glasses but cannot see his eyes. The man can still see through the glasses.

Grade 6-7

2.1 Describe what happens to the light as it passes from the surroundings to the mirrored sunglasses.

....

....

[2]

2.2 One of the lenses of the sunglasses has been damaged, and has become covered in tiny scratches.
Describe how the reflective properties of the damaged and undamaged lenses are different.
State what effect this will have on any reflected images formed by the lenses.

....

....

....

[3]

[Total 5 marks]

Exam Practice Tip

Remember that the angles of incidence and reflection are both measured from the normal — a line perpendicular to the
surface at the point of reflection. Get into the habit of drawing the normal whenever a ray reaches a surface to make
sure you're getting your ray diagrams right. Draw them as dotted lines so that they're not confused with rays.

Electromagnetic Waves and Refraction

1 Electromagnetic waves form a continuous spectrum.

1.1 Use words from the box below to complete the following sentences.

a vacuum	glass	sound	longitudinal	transverse	water

All waves in the electromagnetic spectrum are ....

All electromagnetic waves travel at the same speed in ....

[2]

1.2 **Figure 1** shows a graph of intensity against wavelength for two objects at different temperatures.

Name the part of the electromagnetic spectrum that the peak wavelength of object B lies in.

....

[1]

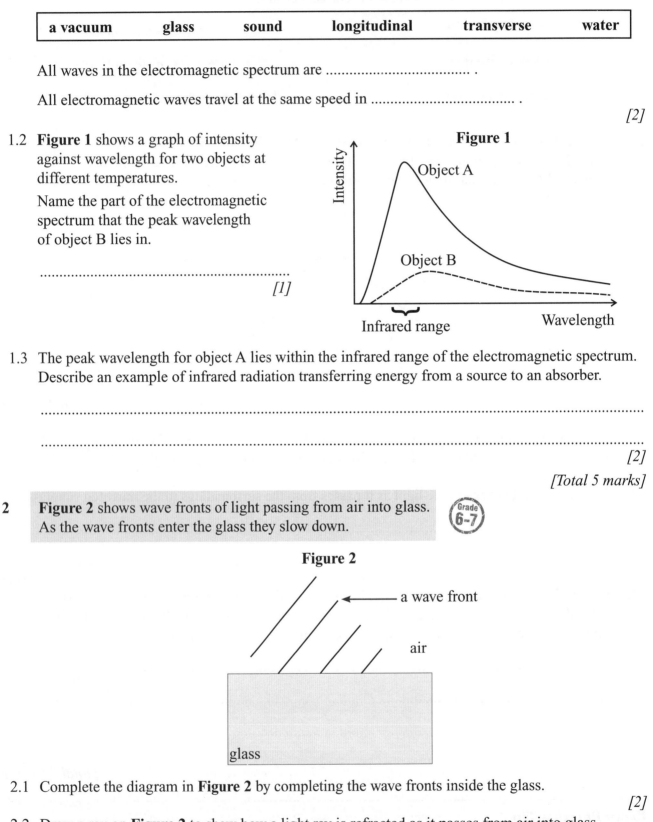

Figure 1

1.3 The peak wavelength for object A lies within the infrared range of the electromagnetic spectrum. Describe an example of infrared radiation transferring energy from a source to an absorber.

....

....

[2]

[Total 5 marks]

2 **Figure 2** shows wave fronts of light passing from air into glass. As the wave fronts enter the glass they slow down.

Figure 2

2.1 Complete the diagram in **Figure 2** by completing the wave fronts inside the glass.

[2]

2.2 Draw a ray on **Figure 2** to show how a light ray is refracted as it passes from air into glass. On your diagram, label the incident ray, refracted ray, and normal line.

[3]

[Total 5 marks]

Topic 6 — Waves

Investigating Light

1   A student was investigating the reflection of light by different types of surface.
     She set up a ray box, a mirror and a piece of white card as shown in **Figure 1**.

Figure 1

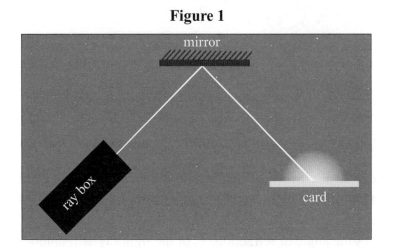

1.1 The student measured the angle of incidence and angle of reflection at the mirror.
     State what she would notice about these measurements.

     ....

     [1]

1.2 Name the type of reflection which is observed at the mirror and at the card.

     At the mirror: ....

     At the card: ....

     [2]

1.3 The student concluded that the surface of the white card was rough, rather than smooth.
     Explain how light is reflected from a rough surface and how this led to the behaviour
     observed at the card.

     ....

     ....

     ....

     [2]

1.4 Explain why a ray-box was used for this experiment. Suggest another piece of apparatus which
     the student could have used to achieve the same effect.

     ....

     ....

     ....

     [3]

     [Total 8 marks]

Topic 6 — Waves

2 A student is investigating refraction through different materials. The student uses a ray box to shine a ray of light into blocks of materials at a fixed angle, *I*. He traces the path of the ray entering and leaving the block on a sheet of paper.

Figure 2

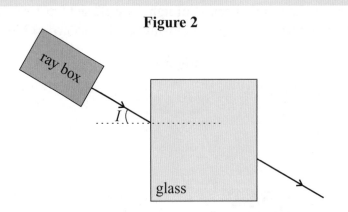

2.1 **Figure 2** shows the student's investigation for light refracted through a glass block. Complete the diagram by drawing the light ray as it passes through the glass block.

[1]

2.2 The student measures the angle of refraction, *R*, of the light ray as it enters the block. **Table 1** shows the results for a range of materials. Measure the angle of refraction for the light ray entering the glass block in **Figure 2**, and hence complete **Table 1**.

[1]

Table 1

Material	*I*	*R*
Cooking Oil	30°	20°
Water	30°	22°
Plastic	30°	20°
Glass	30°	....

2.3 State how the speed of the light ray changes as it passes from air into the glass block. Explain how the result in **Table 1** shows this is the case.

....

....

[2]

2.4 Name the material which changed the speed of the light ray the least. Explain your answer.

....

....

....

[3]

2.5 Cooking oil and water are both liquids, so need to be placed within transparent solid containers to be used in this experiment. Explain why the student should ensure that the containers he uses have thin walls.

....

....

[2]

[Total 9 marks]

Radio Waves

Tick the appropriate boxes to sort the radio-wave facts from the fiction.

	True	False
Long-wave radio can be transmitted across long distances.	☐	☐
Long-wave radio uses diffraction to follow the curve of the Earth's surface.	☐	☐
Short-wave radio can only be used over short distances.	☐	☐
Radio waves with very short wavelengths do not travel well through obstacles.	☐	☐

1* Describe how an electrical signal generates a radio wave in a TV signal transmitter. Explain how this radio wave can generate an electrical signal in a distant TV aerial.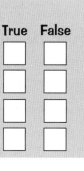

....

....

....

....

....

....

....

[Total 4 marks]

2* A family from northern England are on holiday in France. Explain why they are unable to listen to their local FM radio station from back home, but are still able to listen to the same long-wave radio broadcasts as they do at home.

....

....

....

....

....

....

....

....

....

....

[Total 6 marks]

Topic 6 — Waves

EM Waves and Their Uses

1 A student uses a microwave oven to cook a jacket potato. (Grade 6-7)

1.1 Describe how microwaves cook the potato in the microwave oven.

....

....

[3]

1.2 The potato is placed in the microwave oven on a glass plate.
Explain why the glass plate does not get hot when the microwave oven is used.

....

....

[2]

1.3 Name one other type of electromagnetic radiation which is commonly used to cook food.

....

[1]

1.4 Microwaves can also be used to communicate with satellites. Explain why the microwaves used
for communications must have different wavelengths to those used in microwave ovens.

....

....

....

....

[4]
[Total 10 marks]

2 A police helicopter has an infrared camera attached to its base. The camera
can be used to detect people trying to hide in the dark. Explain the advantages
of using an infrared camera rather than a normal camera for this purpose. (Grade 7-9)

....

....

....

[Total 3 marks]

More Uses of EM Waves

Warm-Up

Sort the EM uses below into the table. Some of the uses may appear in more than one column.

Uses of EM waves

A artificial suntanning

B fibre optic data transmission

C energy efficient light bulbs

D revealing invisible ink

E medical imaging of bones

F cancer treatment

UV Rays	Visible Light	X-rays	Gamma Rays

1 X-rays and gamma rays can both be used in medical imaging. *Grade 6-7*

1.1 Briefly describe how a medical tracer can be used to create an internal body image.

....

....

....

[2]

1.2 Explain why gamma rays are suitable for medical imaging.

....

....

[1]

1.3 Explain how X-rays are used to form images of a patient's skeleton.

....

....

....

....

[3]

1.4 Exposure to both X-rays and gamma rays can be dangerous to humans. Suggest **one** precaution taken by medical workers who use X-rays or gamma rays when imaging patients.

....

[1]

[Total 7 marks]

Exam Practice Tip

You may be asked to explain why a given electromagnetic wave is suited to a particular use. So make sure you understand the properties of the different electromagnetic wave types, and know some of their most common uses.

Dangers of Electromagnetic Waves

1 Some types of electromagnetic wave can be harmful to people.

1.1 Describe how X-rays and gamma rays can cause cancer.

....

....

[2]

1.2 Another type of harmful electromagnetic radiation is ultraviolet radiation.
Give **two** damaging effects of ultraviolet light.

....

....

[2]

[Total 4 marks]

2 **Table 1** lists the radiation doses for some common medical procedures.

Table 1

Procedure	Typical effective dose (mSv)	Lifetime additional risk of fatal cancer per examination
X-ray image of skull	0.07	1 in 300 000
X-ray image of lower spine	1.3	1 in 15 000
CT scan of head	2	1 in 10 000

2.1 The lifetime additional risk of fatal cancer from a CT scan of the chest is 1 in 2500.
Use **Table 1** to estimate the typical effective dose of a CT scan of the chest.

Typical effective dose = ... mSv

[2]

2.2* Nuclear medicine scans use gamma rays to create images of internal organs, but have a high
effective radiation dosage. Discuss the risks involved in performing this type of scan, and why
the procedure might go ahead despite the risks.

....

....

....

....

....

....

....

[6]

[Total 8 marks]

Lenses

1 **Figure 1** shows a lens being used to focus light. The diagram is to scale. **Grade 4-6**

Figure 1

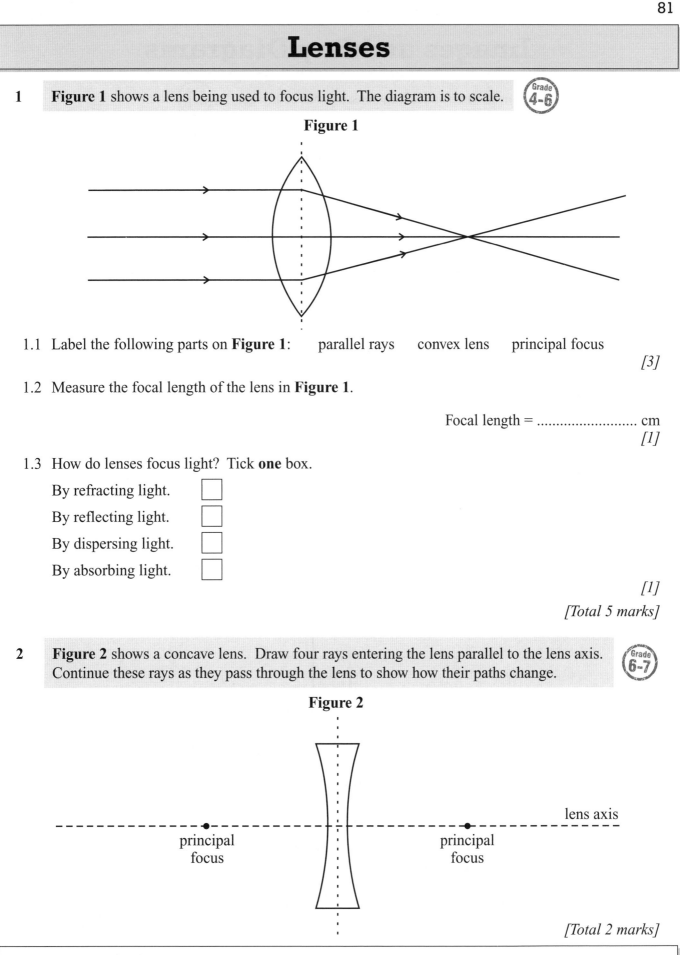

1.1 Label the following parts on **Figure 1**: parallel rays   convex lens   principal focus

[3]

1.2 Measure the focal length of the lens in **Figure 1**.

Focal length = ... cm

[1]

1.3 How do lenses focus light? Tick **one** box.

By refracting light. ☐

By reflecting light. ☐

By dispersing light. ☐

By absorbing light. ☐

[1]

[Total 5 marks]

2 **Figure 2** shows a concave lens. Draw four rays entering the lens parallel to the lens axis. Continue these rays as they pass through the lens to show how their paths change. **Grade 6-7**

Figure 2

principal focus   principal focus   lens axis

[Total 2 marks]

Exam Practice Tip

When drawing ray diagrams with lenses, always make sure you draw the ray which goes through the centre of the lens. This ray will always hit the lens perpendicular to the surface, so it won't be refracted. It can be a useful guide for the rest of the diagram — especially when you're drawing diagrams of images, which you'll encounter on the next page.

Images and Ray Diagrams

1   Ray diagrams are used to visualise how lenses produce images.   **Grade 4-6**

   1.1   Complete the ray diagram in **Figure 1** below. Draw the image formed.

Figure 1

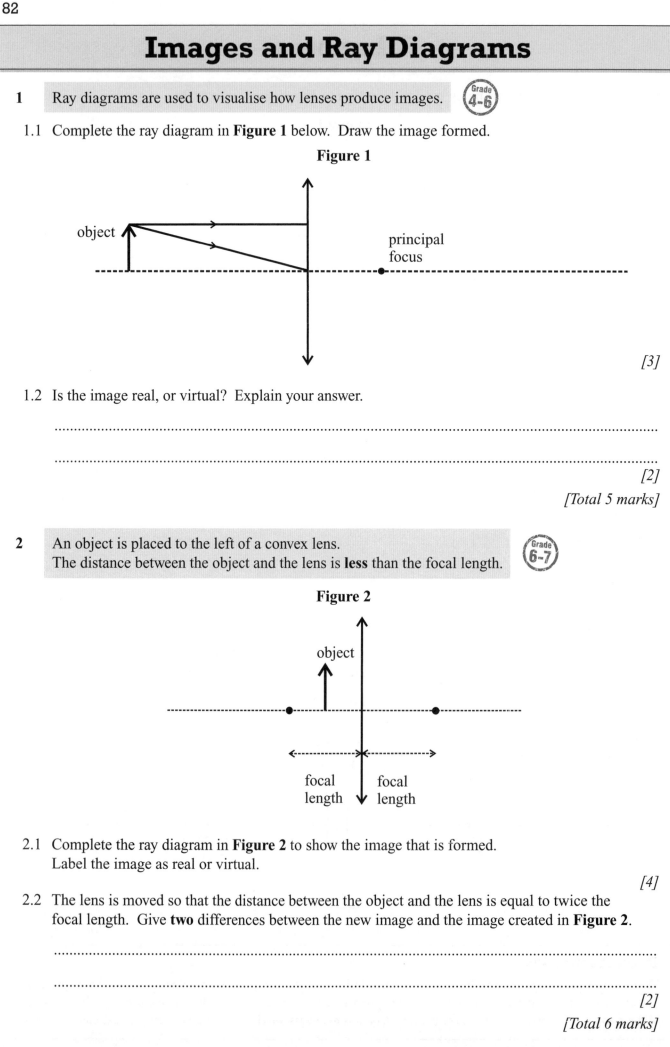

[3]

   1.2   Is the image real, or virtual? Explain your answer.

....

....

[2]

[Total 5 marks]

2   An object is placed to the left of a convex lens.
The distance between the object and the lens is **less** than the focal length.   **Grade 6-7**

Figure 2

   2.1   Complete the ray diagram in **Figure 2** to show the image that is formed.
Label the image as real or virtual.

[4]

   2.2   The lens is moved so that the distance between the object and the lens is equal to twice the
focal length. Give **two** differences between the new image and the image created in **Figure 2**.

....

....

[2]

[Total 6 marks]

Concave Lenses and Magnification

1   **Figure 1** shows some concave lenses.

1.1 Tick the boxes below the **two** diagrams which correctly show how concave lenses refract light.

Figure 1

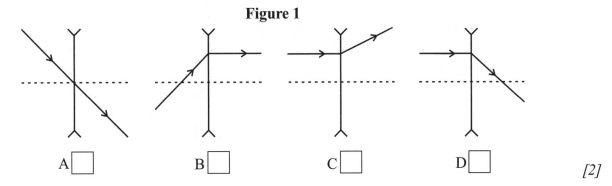

A ☐   B ☐   C ☐   D ☐   *[2]*

1.2 Complete the ray diagram in **Figure 2** to show the image of the object formed by the concave lens.

Figure 2

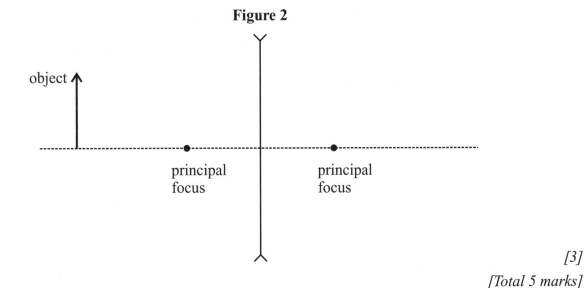

[3]

[Total 5 marks]

2   **Figure 3** shows a convex lens. Complete the ray diagram and calculate the magnification of the lens. The height of the object is 20 mm.

Figure 3

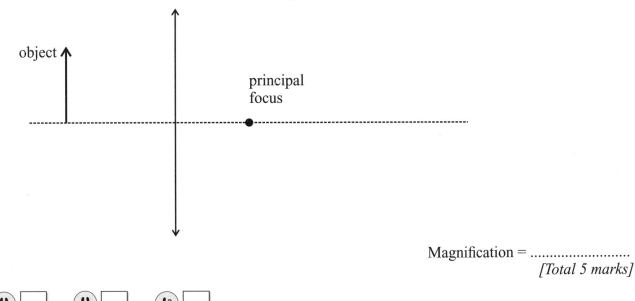

Magnification = ...

[Total 5 marks]

☹ ☐   ☺ ☐   ☺ ☐

Visible Light

Here's a diagram of a transparent green cube.
Continue the appropriate rays to show which colours of light are transmitted by the cube.

Red light →

Green light →

Blue light →

1   A student is investigating colour by looking at two footballs.
One football is red, the other is white.

Grade 4-6

1.1 Both of the footballs are opaque. Describe what is meant by the term opaque.

....

....

[1]

1.2 Describe how the footballs interact with the light waves to make one football look red
and the other football appear white.

....

....

....

[2]

1.3 The student looks at the footballs through a red filter. Both footballs now look red.
Explain why this happens.

....

....

[2]

1.4 The student now looks at the footballs through a green filter.
State and explain what colour each football appears.

....

....

....

....

[3]

[Total 8 marks]

2 **Table 1** shows some wavelengths of light from the visible light spectrum and their corresponding colours.

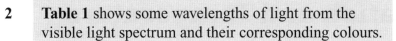

Table 1

Colour	Wavelength (nm)
blue	470
green	540
red	680

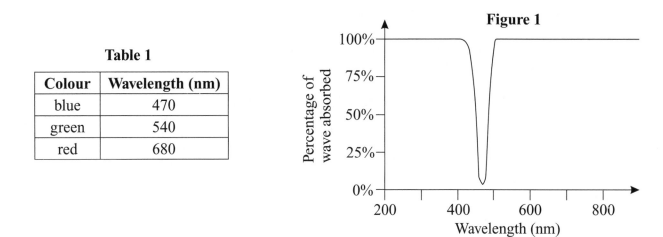

2.1 **Figure 1** shows the percentage of light of different wavelengths absorbed by an object. State whether the object is opaque or transparent.

....

[1]

2.2 What colour is the object?

....

[1]

2.3 Draw the absorption graph on the axes below for a perfectly black object.

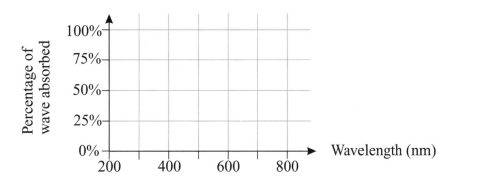

[1]

2.4 A purple colour can be produced from a mixture of red and blue light. On the axes below, sketch the absorption graph for an opaque purple object that reflects only red and blue light.

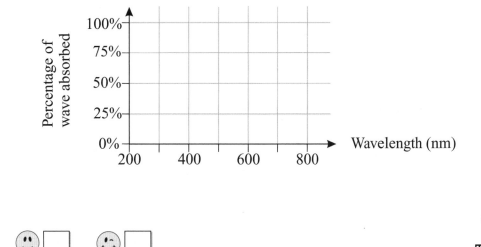

[3]

[Total 6 marks]

Topic 6 — Waves

Infrared Radiation and Temperature

1 All bodies emit and absorb radiation. (Grade 4-6)

1.1 Which of the following statements is true? Tick **one** box.

Only objects that are hotter than 100 °C emit infrared radiation. ☐

Only objects that are cooler than 0 °C absorb infrared radiation. ☐

All objects emit and absorb infrared radiation. ☐

No object can absorb infrared radiation. ☐

[1]

1.2 An object is absorbing the same amount of radiation as it emits.
State what is happening to the temperature of this object.

....

[1]

1.3 State what would happen to the temperature of the object
if it were to emit more radiation than it absorbed.

....

[1]

[Total 3 marks]

PRACTICAL

2 A student uses a Leslie cube, shown in **Figure 1**, to investigate how different materials radiate energy. (Grade 6-7)
A Leslie cube is a hollow cube whose faces are made out of different materials.

Figure 1

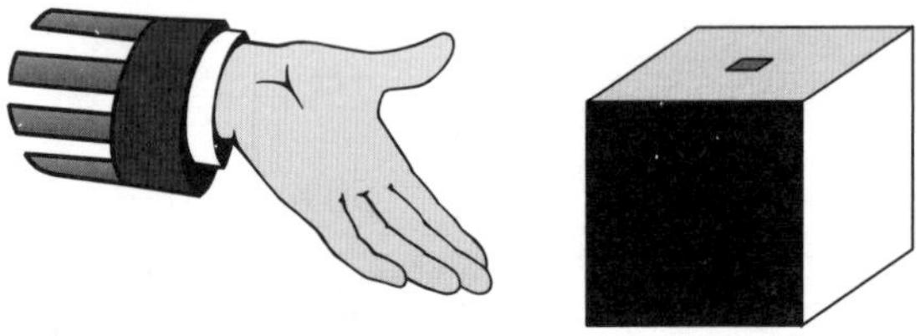

The student fills the cube with hot water and places his hand near to each surface.
He records how warm his hand feels in front of each surface.
The four sides of the cube are matte black, shiny black, matte white and shiny white.

2.1 Predict which side the student's hand would feel warmest in front of.

....

[1]

2.2 Predict which side the student's hand would feel coolest in front of.

....

[1]

2.3 Suggest **one** way to improve the student's experiment.

....

[1]

[Total 3 marks]

☹ ☐   ☺ ☐   ☺ ☐

Black Body Radiation

1 Perfect black bodies are the best possible emitters of radiation. *(Grade 4-6)*

1.1 Define a perfect black body.

....

[1]

1.2 A white star expands and its surface appears redder in colour.
Describe how the surface temperature of the star has changed.

....

[1]

1.3 Which of the following is true for the radiation emitted by any object? Tick **one** box.

The radiation emitted covers a narrow range of wavelengths. ☐

The intensity of radiation emitted is the same for all wavelengths. ☐

The radiation emitted covers a large range of wavelengths. ☐

The intensity of radiation emitted is independent of the object's temperature. ☐

[2]

[Total 4 marks]

2* Electromagnetic radiation from the Sun affects the temperature of the Earth.
Explain, with respect to the radiation emitted by the Sun and the Earth,
how the temperature of the Earth remains approximately constant. *(Grade 7-9)*

....

....

....

....

....

....

....

....

....

[Total 6 marks]

Exam Practice Tip
Remember — the intensity of radiation is just how much energy is being transmitted by the radiation in a certain amount of time. So the more infrared radiation given out by an object, the higher the intensity of infrared radiation.

☹ ☐ 😐 ☐ 🙂 ☐

88

Sound Waves

1 The human ear is specialised to detect sound waves. (Grade 4-6)

1.1 Give the frequency range of normal human hearing.

....

[1]

1.2 Describe the function of the ear drum in the ear.

....

....

[2]

[Total 3 marks]

2 Two children have made a toy telephone out of plastic pots and a piece of string. The string is tied to the bases of the pots and is pulled tight. A child speaking into one plastic pot can be heard by the child at the other end. (Grade 6-7)

2.1 Describe how the toy telephone transmits sounds between the two children.
Your answer should refer to the movement of particles as the sound wave is transmitted.

....

....

....

....

....

....

....

....

....

....

[5]

2.2 State what usually happens to the wavelength and frequency
of a sound wave as it passes from air to a solid material.

....

....

[2]

[Total 7 marks]

Exam Practice Tip

Remember that sound waves come from vibrating objects. They're transmitted through a medium as longitudinal waves.

Topic 6 — Waves

Ultrasound

1 Ultrasound is frequently used in medicine. Grade 4-6

1.1 Define ultrasound.

....

....

[1]

1.2 Tick the boxes next to the frequencies which would produce ultrasound.

☐ 30 mHz   ☐ 30 kHz   ☐ 30 Hz   ☐ 30 MHz

[2]

[Total 3 marks]

2 Describe how ultrasound can be used to create an image of a foetus. Grade 6-7

....

....

....

....

[Total 3 marks]

3 A submarine uses echo-location to work out how far it is above the ocean floor.
It sends a pulse of ultrasound and measures the reflection, as shown in **Figure 1**.
A trace of the original and reflected pulses is shown in **Figure 2**. Grade 6-7

Figure 1

Figure 2

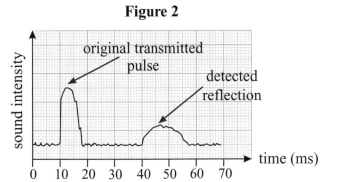

3.1 What is the time interval between the start of the original pulse and
the start of the reflected pulse, as shown in **Figure 2**?

.... ms

[1]

3.2 The sound waves travelled at a speed of 1500 m/s.
Calculate the distance between the submarine and the ocean floor.

....

[3]

[Total 4 marks]

Topic 6 — Waves

Exploring Structures Using Waves

Complete the table shown by stating whether each property describes an S-wave, a P-wave, or both. The first has been done for you.

Property of the wave	Type of wave
Can transfer energy through the Earth	Both
Is a transverse wave	
Can pass through a liquid	
Can pass through a solid	

1   **Figure 1** shows a picture of the Earth. Seismic waves are being produced at point A. Two seismometers are detecting the waves at positions B and C. At B, both S and P waves are detected, but the detector at C is only detecting P waves.

Figure 1

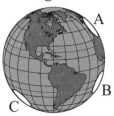

1.1   Explain how this indicates that part of the Earth's core is made from liquid.

....

....

....

[2]

1.2   **Figure 2** shows how the velocity of a P-wave changes as it travels through the Earth.

Figure 2

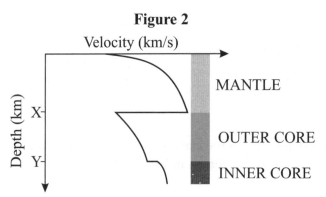

Explain why the velocity of the wave suddenly changes at depth X and Y.

....

....

[2]

[Total 6 marks]

Permanent and Induced Magnets

Complete the sentence using one of the words below.

non-contact   contact   nuclear

Magnetic force is an example of a .... force.

1   Magnets have magnetic fields.   Grade 4-6

1.1 Define the term magnetic field.

....

....

[1]

1.2 Name **two** magnetic materials.

1. ....

2. ....

[2]

1.3 **Figure 1** shows a bar magnet. Draw the magnetic field lines onto the diagram in **Figure 1**.

Figure 1

[2]

1.4 Which of the following statements are correct for magnets? Tick **two** boxes.

Like poles attract each other. ☐

The magnetic field at the north pole of a magnet is always stronger than at the south pole. ☐

The closer together the magnetic field lines, the stronger the magnetic field. ☐

Magnetic field lines point from the north pole to the south pole of a magnet. ☐

The force between a magnet and a magnetic material can be attractive or repulsive. ☐

[2]

[Total 7 marks]

2   A block of cobalt is held in place near to a bar magnet, as shown in **Figure 2**.   (Grade 6-7)

Figure 2

N S	•P
bar magnet	cobalt

2.1   A steel paperclip is placed against the block of cobalt at point P, shown on **Figure 2**.
The paperclip sticks to the block of cobalt. State why this is the case.

....

....

....

[2]

2.2   The bar magnet is removed. Explain what happens to the paperclip.

....

....

[2]

[Total 4 marks]

3   A student wants to investigate the magnetic field of a horseshoe magnet, shown in **Figure 3**.   (Grade 6-7)

Figure 3

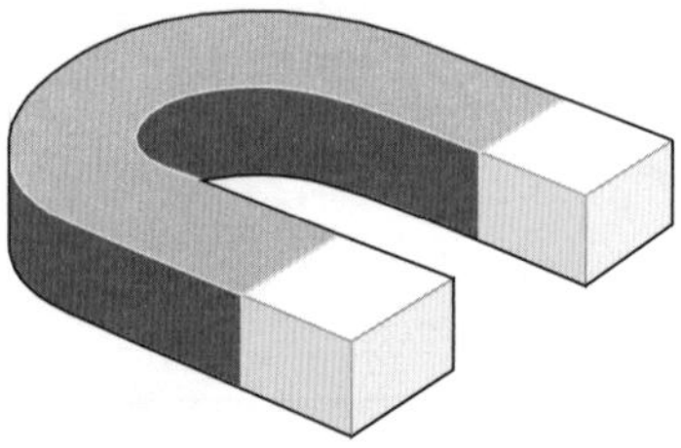

3.1* Explain how a compass could be used to determine the magnetic field of the magnet.

....

....

....

....

....

....

[4]

3.2   State what would happen to the compass if you were to move it far away from any magnets.
Explain why this would happen.

....

....

[2]

[Total 6 marks]

Topic 7 — Magnetism and Electromagnetism

Electromagnetism

1   **Figure 1** shows a wire which has a current flowing through it.
     The arrow show the direction of the current.

Figure 1

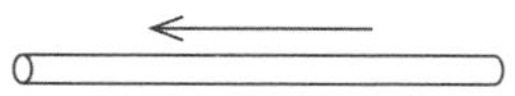

1.1 The flow of charge creates a magnetic field around the wire.
     On **Figure 1**, draw field lines showing the direction of the magnetic field created.

[2]

1.2 The direction of the current is reversed. State the effect this will have on the magnetic field.

....

[1]

1.3 State **one** way in which the magnetic field strength around the wire could be increased.

....

[1]

[Total 4 marks]

2   A solenoid with an iron core is an electromagnet.

2.1 State one difference between permanent magnets and electromagnets.

....

....

[1]

2.2 Describe the magnetic field inside the centre of a solenoid.

....

....

[2]

2.3 Which of the following statements about a current-carrying solenoid is true? Tick **one** box.

There will be no magnetic field if the wire of the solenoid is stretched out straight. ☐

It is not possible to change the strength of the magnetic field around the solenoid. ☐

If the current is stopped, there will no longer be a magnetic field around the solenoid. ☐

The direction of the magnetic field does not depend on the direction of the current. ☐

[1]

[Total 4 marks]

Topic 7 — Magnetism and Electromagnetism

3   A current-carrying solenoid has a magnetic field outside it similar to a bar magnet.   (Grade 7-9)

3.1 State **one** way in which the magnetic field strength of a solenoid can be increased.

....

[1]

3.2 The north pole of a magnet is brought near to the current-carrying solenoid as shown in **Figure 2**.
State whether the north pole is **attracted** or **repelled** by the solenoid. Explain why.

Figure 2

....

....

....

[3]

[Total 4 marks]

4*   **Figure 3** shows a circuit for an electric bell.   (Grade 7-9)

Figure 3

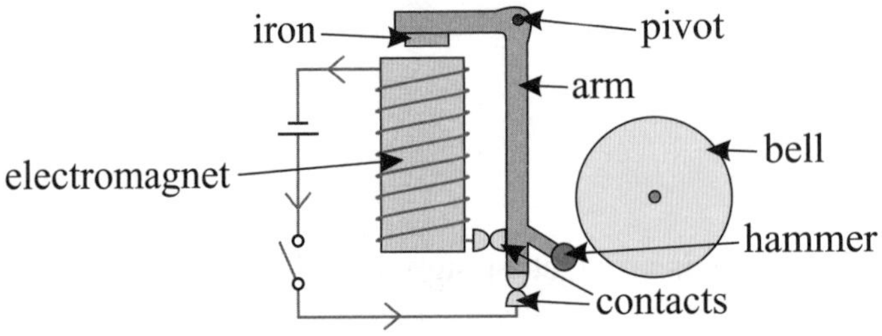

Explain how the circuit uses electromagnetism to sound the bell.

....

....

....

....

....

....

....

....

[Total 6 marks]

Exam Practice Tip

A current carrying-wire will always produce a magnetic field around it. No matter what position the wire is in, or what shape it's been bent into, the magnetic field around it will always depend on the direction of the current.

Topic 7 — Magnetism and Electromagnetism

The Motor Effect

1 A wire is placed between two magnets, as shown in **Figure 1**.   (Grade 4-6)
A current is flowing through the wire, in the direction shown.

Figure 1

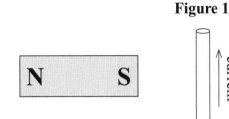

1.1 What will happen to the wire? Tick **one** box.

It will move to the left. ☐

It will move away from you, into the paper. ☐

It will move towards you, out of the paper. ☐

It will remain stationary. ☐

[1]

1.2 State the name of this effect and explain what causes it.

....

....

[2]

1.3 State **three** factors which determine the magnitude of the force acting on the wire.

1. ....

2. ....

3. ....

[3]

[Total 6 marks]

2 A 0.75 m section of wire, carrying a current of 0.4 A, is placed into a magnetic field.   (Grade 6-7)
When the wire is perpendicular to the field, it experiences a force of 1.2 N.
Calculate the magnetic flux density of the field. Give the correct unit in your answer.

Magnetic flux density = ...

Unit = ...

[Total 4 marks]

Exam Practice Tip

If you're struggling to remember Fleming's left-hand rule, think thu**M**b, **F**irst finger and se**C**ond finger.

☹ ☐   😐 ☐   🙂 ☐   Topic 7 — Magnetism and Electromagnetism

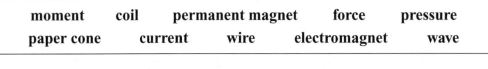

Electric Motors and Loudspeakers

1   The loudspeaker in a pair of headphones uses an alternating current to produce sound.
     Use words from the box to complete the description of how loudspeakers work.

| moment | coil | permanent magnet | force | pressure |
| paper cone | current | wire | electromagnet | wave |

An alternating current is passed through a coil of wire. The coil is surrounded by a

.... and attached to the base of a paper cone. When the coil carries

a current, it experiences a ...., so the paper cone moves.

This allows variations in .... to be converted into variations

in .... in sound waves.

[Total 3 marks]

2   **Figure 1** shows part of a basic dc motor. A coil of wire is
     positioned between two magnetic poles and allowed to rotate.

Figure 1

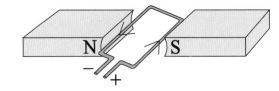

2.1 State the direction in which the coil will turn (**anticlockwise** or **clockwise**).

     ....
                                                                                                                   [1]

2.2 Explain why the coil turns.

     ....

     ....

     ....
                                                                                                                   [2]

2.3 Explain how a dc current can be used to make the coil
     in **Figure 1** continue to rotate in the same direction.

     ....

     ....

     ....

     ....
                                                                                                                   [2]
                                                                                                        [Total 5 marks]

The Generator Effect

1 This question is about statements A and B, shown below. *(Grade 6-7)*

> **A** A potential difference is induced when an electrical conductor moves relative to a magnetic field.
>
> **B** A potential difference is induced when there is a change in the magnetic field around an electrical conductor.

Which of the following is correct? Tick **one** box.

Only statement **A** is true. ☐

Only statement **B** is true. ☐

Both statements **A** and **B** are true. ☐

Neither statement **A** nor **B** is true. ☐

[Total 1 mark]

2 **Figure 1** shows a device fixed to the frame of a bicycle. When the bicycle wheel turns, it causes the generator wheel to turn. This then generates electricity to power the bicycle's lamp. *(Grade 7-9)*

Figure 1

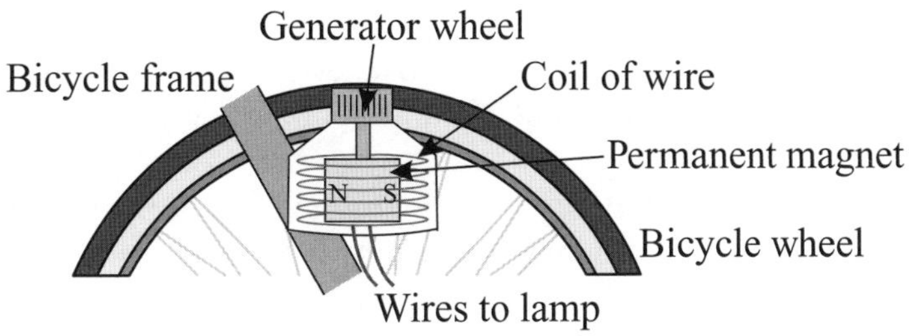

Generator wheel

Bicycle frame — Coil of wire

— Permanent magnet

N S

Bicycle wheel

Wires to lamp

2.1 Describe how the device uses the generator effect to power the bicycle's lamp.

....

....

....

....

[3]

2.2 A cyclist is riding a bicycle with the device fitted. She wants to increase the lamp's brightness. State **one** way in which she can do this.

....

[1]

2.3 The current that the device generates creates its own magnetic field. Describe the direction that the magnetic field will be created in.

....

....

[1]

[Total 5 marks]

☹ ☐   ☺ ☐   😉 ☐

Generators and Microphones

Draw lines between the boxes to describe each type of generator.

Alternators...

Dynamos...

...use a split-ring commutator...

...use slip rings and brushes...

...and generate dc.

...and generate ac.

1   **Figure 1** shows part of the inside of a microphone. *Grade 6-7*

Figure 1

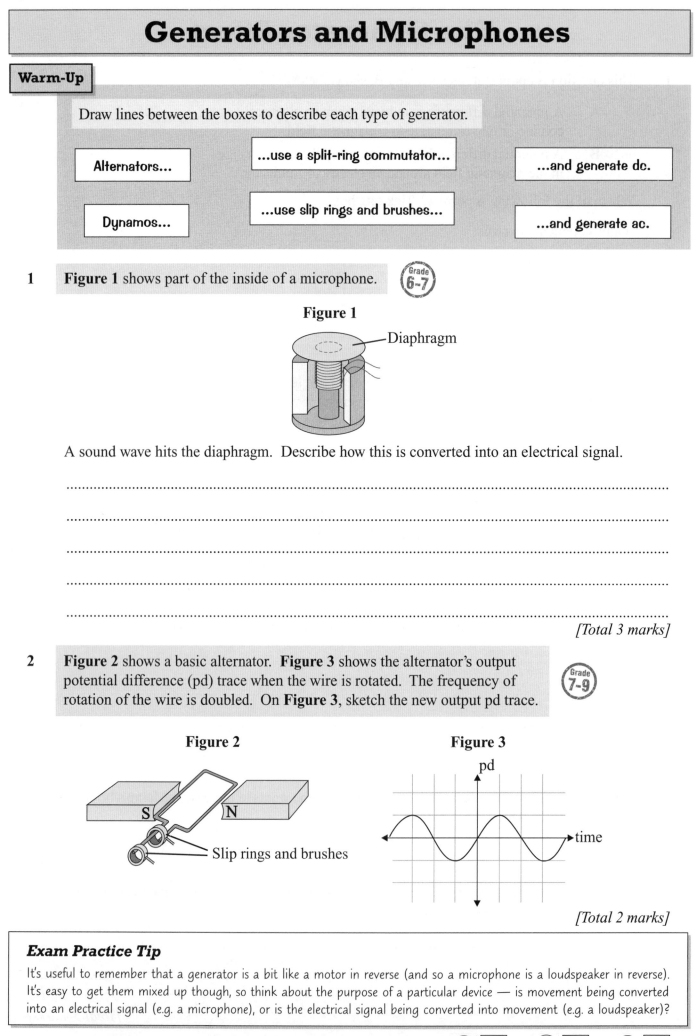

Diaphragm

A sound wave hits the diaphragm. Describe how this is converted into an electrical signal.

....

....

....

....

....

[Total 3 marks]

2   **Figure 2** shows a basic alternator. **Figure 3** shows the alternator's output potential difference (pd) trace when the wire is rotated. The frequency of rotation of the wire is doubled. On **Figure 3**, sketch the new output pd trace. *Grade 7-9*

Figure 2

S   N

Slip rings and brushes

Figure 3

pd

time

[Total 2 marks]

Exam Practice Tip

It's useful to remember that a generator is a bit like a motor in reverse (and so a microphone is a loudspeaker in reverse). It's easy to get them mixed up though, so think about the purpose of a particular device — is movement being converted into an electrical signal (e.g. a microphone), or is the electrical signal being converted into movement (e.g. a loudspeaker)?

Topic 7 — Magnetism and Electromagnetism   ☹ ☐   ☺ ☐   ☺ ☐

Transformers

1 Put these statements in the correct order to describe how current flows in a transformer. *(Grade 4-6)*

    A   This causes a changing magnetic field in the iron core.

    B   If the secondary coil is part of a complete circuit, this causes a current to be induced.

    C   This changing magnetic field induces an alternating pd in the secondary coil.

    D   An alternating current is applied across the primary coil.

Correct order: ....

[Total 1 mark]

2 A transformer has 12 turns in the primary coil. An input potential difference of 240 V is converted to an output potential difference of 80 V. *(Grade 6-7)*

2.1 Calculate the number of turns on the secondary coil.

Number of turns = ...

[3]

2.2 State the type of transformer described above.

....

[1]

[Total 4 marks]

3 A transformer has 30 turns on the primary coil and 40 turns on the secondary coil. The potential difference across the primary coil is 12 V. You can assume that the transformer is 100% efficient. *(Grade 6-7)*

3.1 Calculate the output potential difference.

Output pd = ... V

[3]

3.2 The current and potential difference for the transformer is altered. The current through the primary coil is now 20 A and the potential difference is 30 V. The potential difference across the secondary coil is 40 V. Calculate the current through the secondary coil.

Output current = ... A

[3]

3.3 Suggest **one** reason why the change in 3.2 might be made.

....

[1]

[Total 7 marks]

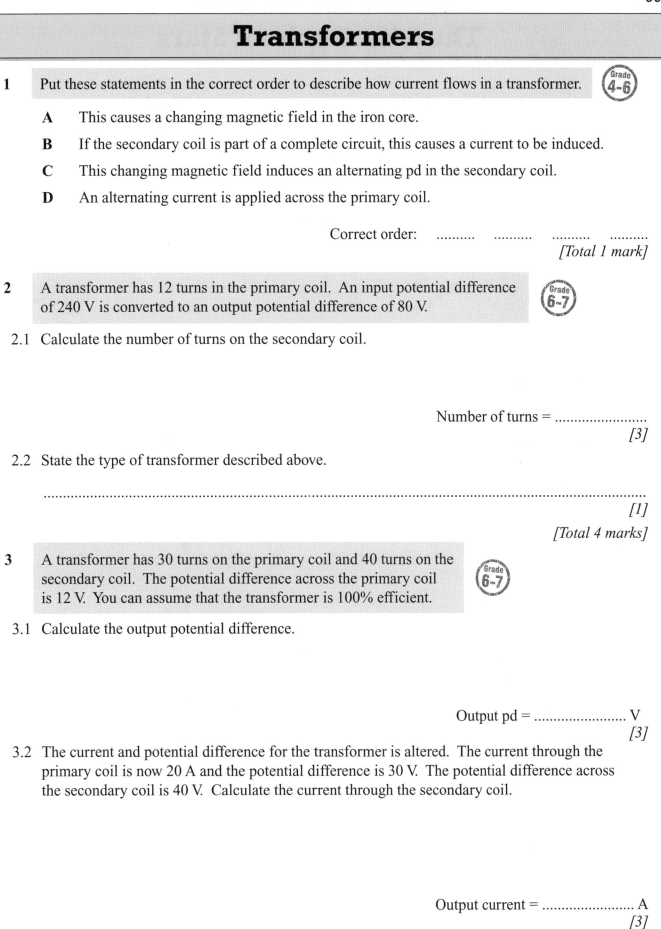

Topic 7 — Magnetism and Electromagnetism

The Life Cycle of Stars

1   A star begins to form when a nebula is pulled together and compressed.

1.1 What is meant by the term 'nebula'?

....

[1]

1.2 State the force responsible for this 'pulling together'.

....

[1]

[Total 2 marks]

2   The basic life cycle of a star with the same mass as the Sun is shown below.
Fill in the gaps to complete the life cycle of a star of 25 times the mass of the Sun.

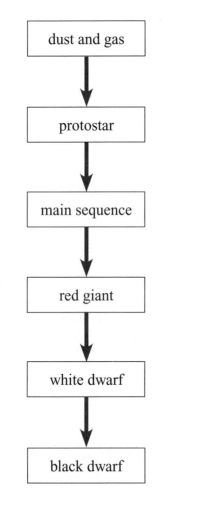

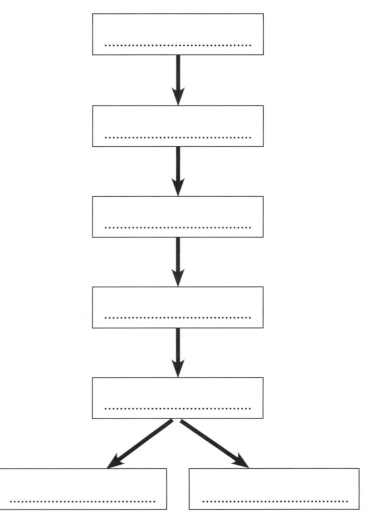

[Total 5 marks]

3 Stars are fuelled by a process known as nuclear fusion.

3.1 Use the correct words from the box to complete the following passage on nuclear fusion in stars.

density	temperature	hydrogen	nebula
bond	helium	collide	iron

As a protostar ages, its .... and .... increase.

This causes particles to .... with each other more often. When the temperature

gets hot enough, ... nuclei fuse together and create .... nuclei.

This process is known as nuclear fusion.

[3]

3.2 When a star reaches the 'main sequence' stage of its lifetime, it is stable.
Explain how nuclear fusion keeps a star stable.

....

....

....

[2]

3.3 State the effect fusion has on the main sequence star's size, and the temperature of its core.

....

....

[2]

[Total 7 marks]

4 Our Sun is currently in the main sequence phase of its life cycle. When the Sun runs out of fuel, it will eject its outer layers to form a planetary nebula, leaving behind a white dwarf, which will gradually cool and fade away. Describe the final life cycle stages of a star with a mass much greater than our Sun, once it has stopped being a red supergiant.

....

....

....

....

....

....

[Total 3 marks]

Exam Practice Tip

This is one of those bits of physics where there's just a lot of words and facts you need to learn. You need to remember the names of all of the different stages of the life cycles of stars of different sizes, as well as what's going on in each stage.

Topic 8 — Space Physics

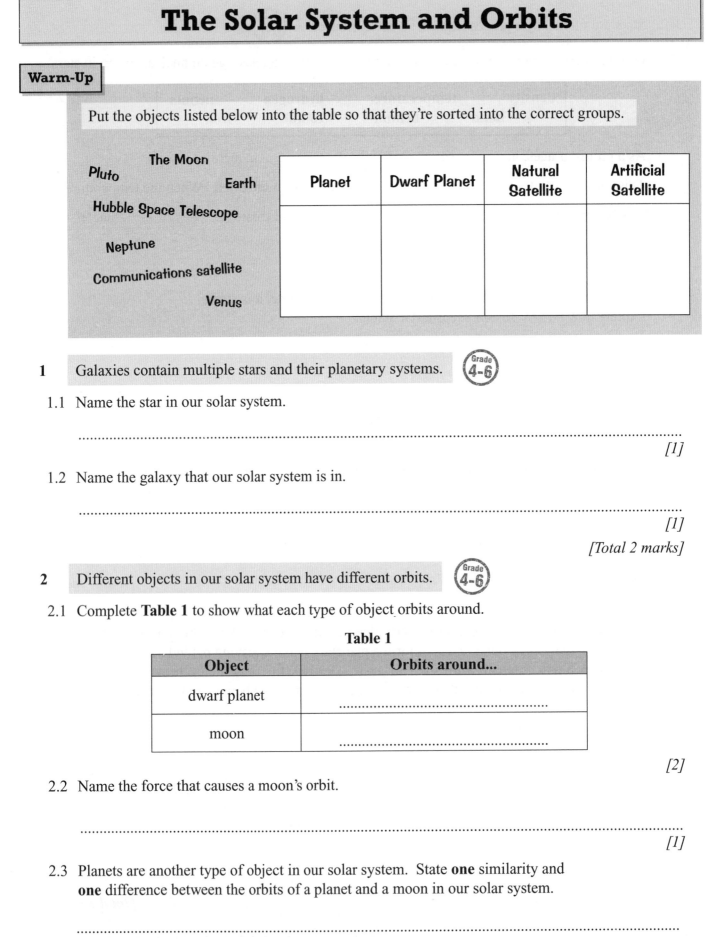

The Solar System and Orbits

Warm-Up

Put the objects listed below into the table so that they're sorted into the correct groups.

The Moon
Pluto
Earth
Hubble Space Telescope
Neptune
Communications satellite
Venus

Planet	Dwarf Planet	Natural Satellite	Artificial Satellite

1 Galaxies contain multiple stars and their planetary systems. (Grade 4-6)

1.1 Name the star in our solar system.

....

[1]

1.2 Name the galaxy that our solar system is in.

....

[1]

[Total 2 marks]

2 Different objects in our solar system have different orbits. (Grade 4-6)

2.1 Complete **Table 1** to show what each type of object orbits around.

Table 1

Object	Orbits around...
dwarf planet	....
moon	....

[2]

2.2 Name the force that causes a moon's orbit.

....

[1]

2.3 Planets are another type of object in our solar system. State **one** similarity and **one** difference between the orbits of a planet and a moon in our solar system.

....

....

[2]

[Total 5 marks]

3 The planets all move in circular orbits around the Sun. They each experience a constant acceleration.

3.1 On **Figure 1**, draw arrows indicating the direction of the planet's acceleration, and the direction of the planet's instantaneous velocity.

Figure 1

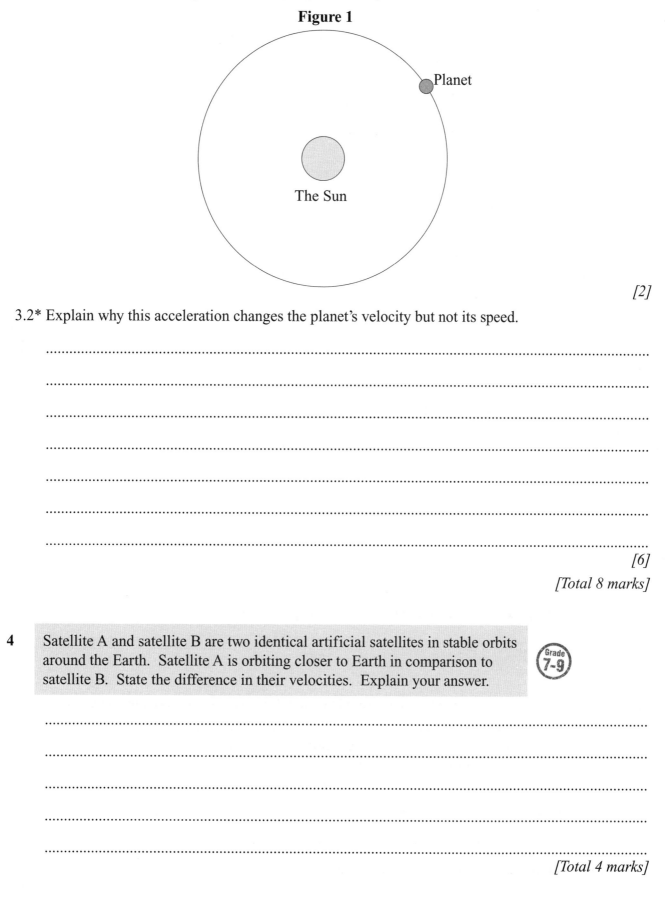

[2]

3.2* Explain why this acceleration changes the planet's velocity but not its speed.

....

....

....

....

....

....

....

[6]

[Total 8 marks]

4 Satellite A and satellite B are two identical artificial satellites in stable orbits around the Earth. Satellite A is orbiting closer to Earth in comparison to satellite B. State the difference in their velocities. Explain your answer.

....

....

....

....

....

[Total 4 marks]

Red-shift and the Big Bang

1 The Big Bang theory is the currently-accepted theory of how the universe was formed. **Grade 4-6**

1.1 Which **two** of the following statements form the basis of the Big Bang theory?
Put ticks in the appropriate boxes to indicate your answers.

[2]

☐ The universe will one day collapse.

☐ The universe started off hot and dense.

☐ The universe is expanding.

☐ The universe has existed forever.

1.2 Many things about the universe are still not understood.
Name **one** feature of the universe that is still unexplained.

....

[1]

[Total 3 marks]

2 Many theories of the universe suggest it is expanding. **Grade 6-7**

2.1 Give **two** observations that support the idea that the universe is expanding.

....

....

....

[2]

2.2 Explain why galaxies are not pulled apart by the expansion of the universe.

....

[1]

2.3 Use the correct words from the box to complete the following passage about the universe.

stabilising	accelerating	decelerating
decreased	increased	stabilised

Recent observations of distant supernovae indicate that the speed at which

distant galaxies are receding has ....

This suggests that the expansion of the universe is ....

[2]

[Total 5 marks]

3 **Table 1** shows a list of galaxies and their distance from Earth in light years.

Table 1

Galaxy	Distance From Earth (light years)
Cigar Galaxy	12 million
Black Eye Galaxy	24 million
Sunflower Galaxy	37 million
Tadpole Galaxy	420 million

3.1 Light from distant galaxies is observed to undergo red-shift.
Explain what is meant by the term 'red-shift'.

....

....

[1]

3.2 Suggest which of the galaxies in **Table 1** will have the greatest observed red-shift.
Explain your answer.

....

....

....

[3]

3.3 State how you would expect the observed red-shift of light from the Black Eye Galaxy to
have changed in 2000 years' time compared to its current red-shift. Explain your answer.

....

....

....

[3]

3.4 A new galaxy is discovered. Light from this galaxy is red-shifted more than that of
the Cigar Galaxy, but less than that of the Sunflower Galaxy.
Suggest a possible distance of this new galaxy from the Earth.

....

[1]

[Total 8 marks]

Exam Practice Tip

The Big Bang theory is the leading theory of the creation of the universe — but it wasn't always. In the exam, you might be asked to describe why theories can be changed over time. So be prepared to think about why this theory has become so widely accepted, and what would cause a new theory to replace it as the favoured explanation.

Topic 8 — Space Physics

Mixed Questions

1 A student uses a compass to investigate magnetic field patterns.

1.1 The compass contains a permanent bar magnet.
Describe the difference between a permanent magnet and an induced magnet.

....

....

[2]

1.2 The student moves the compass around the current-carrying solenoid shown in **Figure 1**.
The student uses the compass to plot the magnetic field produced by the solenoid.
Sketch the magnetic field produced by the solenoid on **Figure 1**.

Figure 1

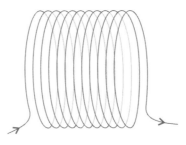

[3]

[Total 5 marks]

2 Background radiation is around us all the time.

2.1 Give **one** man-made source of background radiation.

....

[1]

2.2 Use words from the box below to complete the passage about radioactive decay.
You can only use a word once and you do not need to use all of the words.

can	stable	cannot	unstable	forced	random

Radioactive decay is where a nucleus releases radiation to become more ....

It is a .... process, which means you .... predict

which individual nucleus in a sample will decay next.

[2]

2.3 The term 'activity' can be used when describing a radioactive source.
Define activity and state the unit it is measured in.

....

[2]

2.4 The term 'half-life' can also be used when describing a source of radiation.
Define half-life in terms of activity.

....

[1]

[Total 6 marks]

3 **Figure 2** shows an electric fan.

Figure 2

3.1 The fan is connected to the mains with a cable that contains three wires.
What is the name of this type of cable? Tick **one** box.

☐ three-colour cable   ☐ two-core cable   ☐ three-core cable   ☐ triple cable

[1]

3.2 Complete **Table 1** to show the properties of each wire in the cable.

Table 1

Name of wire	Colour of insulation	Potential difference (V)
Live	....	....
....	blue	....
....	....	0

[3]

3.3 The fan works by transferring energy. Use phrases from the box below to complete the passage.
You can only use a phrase once and you do not need to use all of the phrases.

electrically   thermal   by heating   kinetic   mechanically   elastic potential

Energy is transferred .... from the mains supply to the

.... energy store of the fan's blades.

[2]

3.4 The fan has a power of 30 W. Calculate the energy transferred by the fan in 30 minutes.

Energy transferred = ... J

[2]

[Total 8 marks]

Mixed Questions

4 A student tests the relationship between potential difference and current for a filament bulb. **Grade 4-6**

4.1 **Figure 3** shows four *I-V* characteristics.
Tick the box under the *I-V* characteristic for a filament bulb.

Figure 3

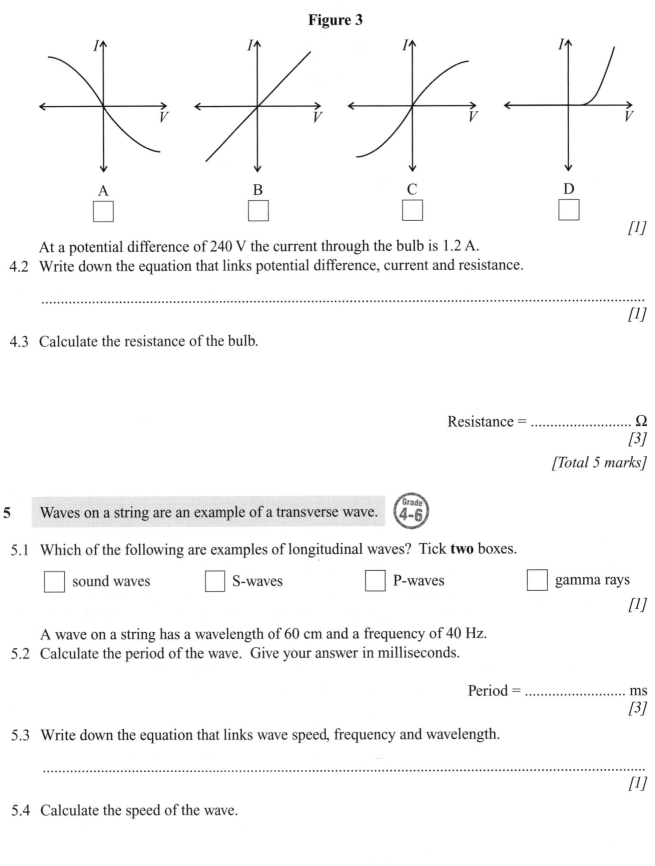

A ☐ B ☐ C ☐ D ☐

[1]

At a potential difference of 240 V the current through the bulb is 1.2 A.

4.2 Write down the equation that links potential difference, current and resistance.

....

[1]

4.3 Calculate the resistance of the bulb.

Resistance = ... Ω

[3]

[Total 5 marks]

5 Waves on a string are an example of a transverse wave. **Grade 4-6**

5.1 Which of the following are examples of longitudinal waves? Tick **two** boxes.

☐ sound waves ☐ S-waves ☐ P-waves ☐ gamma rays

[1]

A wave on a string has a wavelength of 60 cm and a frequency of 40 Hz.

5.2 Calculate the period of the wave. Give your answer in milliseconds.

Period = ... ms

[3]

5.3 Write down the equation that links wave speed, frequency and wavelength.

....

[1]

5.4 Calculate the speed of the wave.

Speed = ... m/s

[2]

[Total 7 marks]

Mixed Questions

6 A child is playing with a remote-controlled toy car. (Grade 6-7)

Figure 4

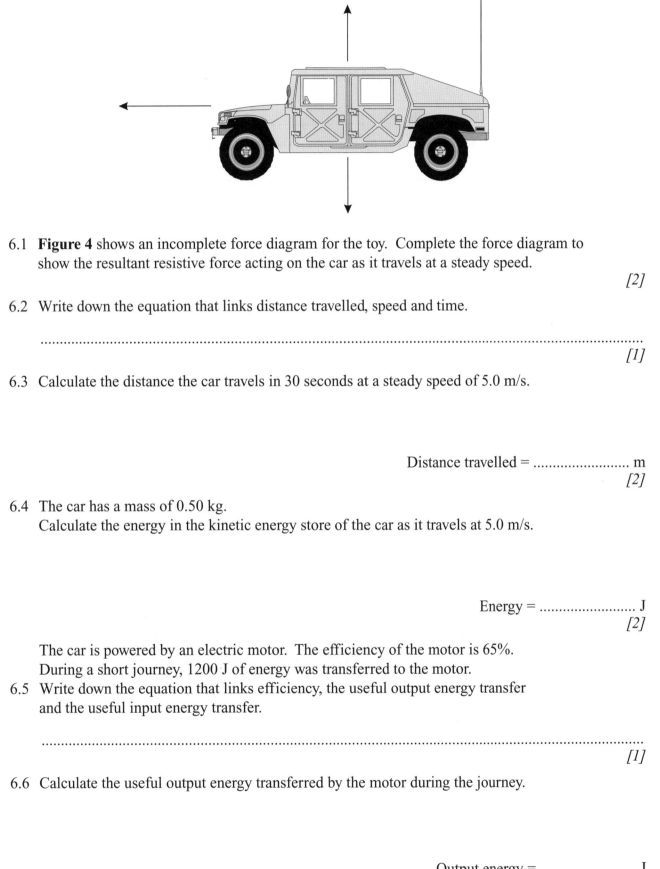

6.1 **Figure 4** shows an incomplete force diagram for the toy. Complete the force diagram to show the resultant resistive force acting on the car as it travels at a steady speed.

[2]

6.2 Write down the equation that links distance travelled, speed and time.

....

[1]

6.3 Calculate the distance the car travels in 30 seconds at a steady speed of 5.0 m/s.

Distance travelled = ... m

[2]

6.4 The car has a mass of 0.50 kg.
Calculate the energy in the kinetic energy store of the car as it travels at 5.0 m/s.

Energy = ... J

[2]

The car is powered by an electric motor. The efficiency of the motor is 65%.
During a short journey, 1200 J of energy was transferred to the motor.

6.5 Write down the equation that links efficiency, the useful output energy transfer and the useful input energy transfer.

....

[1]

6.6 Calculate the useful output energy transferred by the motor during the journey.

Output energy = ... J

[2]

[Total 10 marks]

Mixed Questions

7   A 10 cm × 10 cm × 10 cm block of material A is placed in a beaker of water. It floats when 7 cm of the cube is submerged in the water, as shown in **Figure 6**. Material A is less dense than water.

Grade 6-7

Figure 5

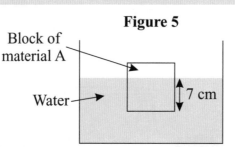

Block of material A

Water

7 cm

7.1   Explain, in terms of the forces acting on the cube, why the cube floats in water.

....

....

....

....

....

[4]

7.2   Write down the equation that links density, mass and volume.

....

[1]

7.3   Water has a density of 1000 kg/m³. Calculate the mass of the water displaced by the cube.

Mass = ... kg

[4]

[Total 9 marks]

8   1.2 kg of water, initially at 10.0 °C, is heated in a saucepan until it boils. The saucepan is left until all of the water evaporates. Calculate how much energy has been transferred to the water during this process. Give your answer in kJ and to 2 significant figures. The specific heat capacity of water is 4.2 kJ/kg. The specific latent heat of vaporisation is 2300 kJ/kg.

Energy transferred = ... kJ

[Total 5 marks]

9   **Figure 6** shows a velocity-time graph for a cyclist's journey.   Grade 6-7

Figure 6

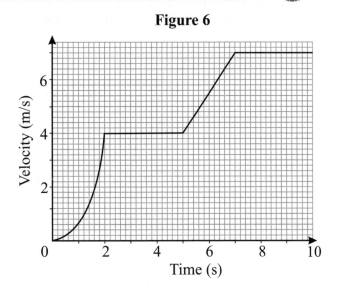

9.1 Describe the motion of the cyclist:

During the first two seconds of the journey: ....

At four seconds from the start of the journey: ....

Between 5-7 s from the start of the journey: ....

[3]

9.2 Calculate the acceleration of the cyclist 6 seconds from the start of his bike ride.

Acceleration = ... m/s^2

[3]

9.3 Ten seconds after the beginning of the cyclist's bike ride, a car turns out of a junction 12 m in front of him. The cyclist is alert and quickly applies the bike's brakes, which provide a constant braking force of 440 N. The combined mass of the man and the bicycle is 83 kg.
Calculate the deceleration of the bicycle.

Deceleration = ... m/s^2

[3]

9.4 Use your answer from 9.3 to determine whether or not the cyclist will hit the car.
Write down any assumptions you make about the cyclist's reaction time.

....

[5]

[Total 14 marks]

Mixed Questions

10 A student is investigating the pressures of different liquids. She fills three identical
containers with a different liquid, then places a pressure sensor in each one. The sensor
is held at the same depth in each case, as shown in **Figure 7**. **Table 2** shows her results.

Figure 7

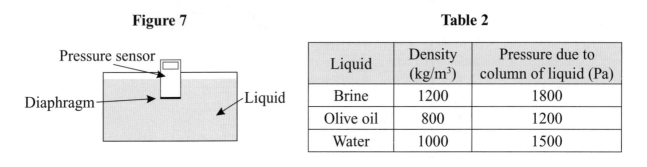

Table 2

Liquid	Density (kg/m³)	Pressure due to column of liquid (Pa)
Brine	1200	1800
Olive oil	800	1200
Water	1000	1500

10.1 Write down the equation that links pressure, force and area.

....

[1]

10.2 The pressure sensor diaphragm has an area of 5.0×10^{-3} m².
Calculate the force exerted on the diaphragm by olive oil in the experiment shown in **Figure 11**.

Force = ... N

[3]

10.3 At a depth of 15 cm, the pressure caused by a fourth liquid is 2850 Pa.
Calculate the density of the new liquid. Use an equation from the Equations List.

Density = ... kg/m³

[3]

10.4* Using the particle model, explain why, at a given depth, the pressure caused by a
column of the new liquid is larger than the pressure caused by a column of water.

....

....

....

....

....

....

....

[4]

[Total 11 marks]

11 A student is investigating the properties of visible light. She uses
the set-up in **Figure 8** to test how different materials refract light.

(Grade 7-9)

She begins by shining a thin ray of white light into one block of material and marking where the
light ray emerges from the block. She then places a block of a different material next to the first
block, leaving no air gap. She repeats this for a range of transparent and translucent materials.

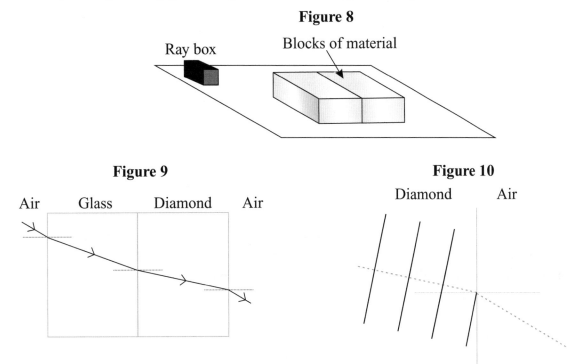

Figure 8

Ray box   Blocks of material

Figure 9   **Figure 10**

Air   Glass   Diamond   Air   Diamond   Air

11.1 Complete the wave front diagram in **Figure 10** for the light ray crossing the boundary between
diamond and air.

[2]

11.2* Explain why the light ray refracts when it crosses the boundary between diamond and air.
Your answer should refer to the wave fronts drawn in **Figure 10**.

....

....

....

....

....

....

....

....

....

[6]

[Total 8 marks]

12   A student is designing a basic electronic toy. He wants the toy to be able to light up and spin around. He creates a basic circuit of a battery connected to a motor. He connects two filament bulbs and a fixed resistor in parallel to the motor. The two bulbs and the resistor are all in series with each other. The bulbs and the motor can be switched on and off separately.

12.1 Draw the circuit diagram for the circuit created by the student.

[5]

12.2 The student turns on the motor alone. The potential difference across the motor is 6.0 V and a current of 70.0 mA flows through the motor. After 15 minutes, the student switches off the motor and measures the temperature of the motor's casing.
He finds that it has increased by 7.0 °C.

The motor's casing has a mass of 25.0 g.
The material it is made from has a specific heat capacity of 120 J/kg °C.
Calculate the amount of energy that is usefully transferred by the motor in 15 minutes.
You can assume that all energy not transferred to thermal energy store of the motor's casing is usefully transferred.

Energy usefully transferred = ... J

[5]

12.3 Explain **one** modification that the student could make to the toy to make it more efficient.

....

....

....

[2]

[Total 12 marks]

Mixed Questions

13 **Figure 11** shows a basic model of how the national grid uses step-up and step-down transformers to vary the potential difference and current of the electricity it transmits.

Grade 7-9

Figure 11

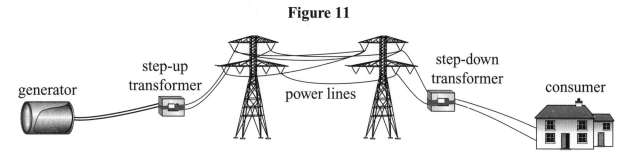

13.1 A power station's generator produces electricity at a potential difference of 25 kV and a current of 4100 A. It is connected to the national grid by a step-up transformer that has 1400 turns on its primary coil and 21 000 turns on its secondary coil. You can assume that the transformer is 100% efficient. Calculate the current of the electricity transmitted by the national grid.
Use equations from the Equations List.

Current = ... A

[4]

13.2 Each minute, 34.92 GJ of energy is transferred to the generator.
Calculate the efficiency of the generator, assuming its power output is constant.

Efficiency = ... %

[5]

13.3 Describe the unwanted energy transfers that occur whilst electricity is being transmitted by the national grid. Explain the causes of these transfers and why transmitting electricity at a lower current reduces these unwanted energy transfers.

....

....

....

....

....

....

....

[5]

[Total 14 marks]

Mixed Questions

14 Nuclear power stations generate electricity from nuclear fission. (Grade 7-9)

14.1* Describe the steps involved in producing a forced nuclear fission reaction in a power station. Explain how control rods, which absorb neutrons, can be used to control the plant's output power.

....

....

....

....

....

....

....

....

....

....

[6]

14.2* Nuclear fission produces nuclear waste. One radioactive isotope in nuclear waste is caesium-137. Caesium-137 has a half-life of 30 years. It produces beta and gamma radiation as it decays. Explain the safety implications of storing the nuclear waste produced by the plant and the precautions needed to reduce the risks posed by storing nuclear waste.

....

....

....

....

....

....

....

....

....

....

[6]

[Total 12 marks]

Exam Practice Tip
Remember to look out for questions like these, where you get marks for how well you write your answer. You could write some short bullet points as a quick plan to help you organise what you want to talk about before writing your full answer.

PAQ41

Mixed Questions